SMP interact

for GCSE

Book I1 part A

I1
PART A

PATHFINDER
EDITION

PUBLISHED BY THE PRESS SYNDICATE OF THE UNIVERSITY OF CAMBRIDGE
The Pitt Building, Trumpington Street, Cambridge, United Kingdom

CAMBRIDGE UNIVERSITY PRESS
The Edinburgh Building, Cambridge CB2 2RU, UK
40 West 20th Street, New York, NY 10011-4211, USA
10 Stamford Road, Oakleigh, VIC 3166, Australia
Ruiz de Alarcón 13, 28014 Madrid, Spain
Dock House, The Waterfront, Cape Town 8001, South Africa

http://www.cambridge.org

© School Mathematics Project 2001
First published 2001

Printed in Italy by Rotolito Lombarda
Typeface Minion   *System* QuarkXPress®

A catalogue record for this book is available from the British Library

ISBN 0 521 01228 7   paperback

Acknowledgements

The authors and publishers are grateful to the following Examination Boards
for permission to reproduce questions from past examination papers:

AQA(NEAB)	Assessment and Qualifications Alliance
AQA(SEG)	Assessment and Qualifications Alliance
Edexcel	Edexcel Foundation
OCR	Oxford, Cambridge and RSA Examinations
WJEC	Welsh Joint Education Committee

1 Parallel lines and angles

A Parallel lines crossing

When one set of parallel lines crosses a different set, you get lots of equal angles.

Sketch this diagram.

Mark with a *p* every angle that equals angle *p*.

Mark with a *q* every angle that equals angle *q*.

- If you know the size of *p*, how do you work out *q*?

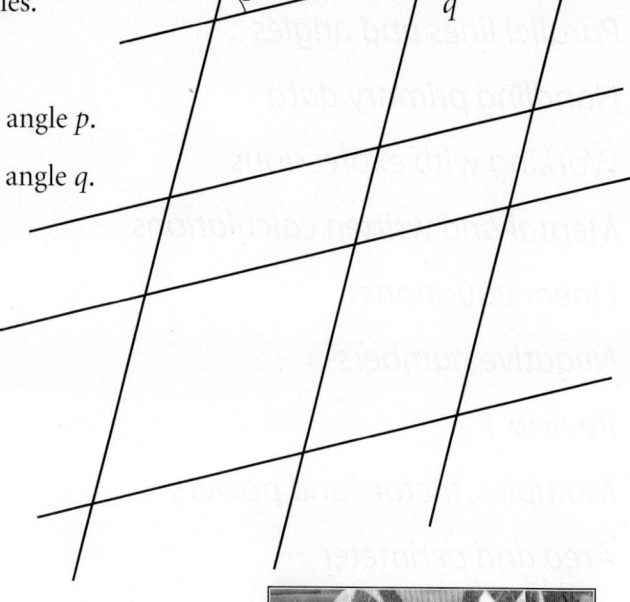

A1 You can buy a wooden trellis from a garden centre. It looks like this when you buy it.

You can 'expand' it and hang it on a garden wall so plants grow up it.

(a) If this angle is 50°, what will the angles marked with letters be?

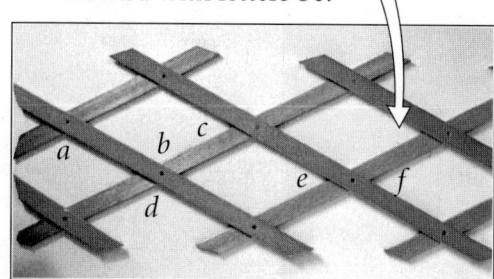

(b) If this angle is 110°, what will the angles marked with letters be?

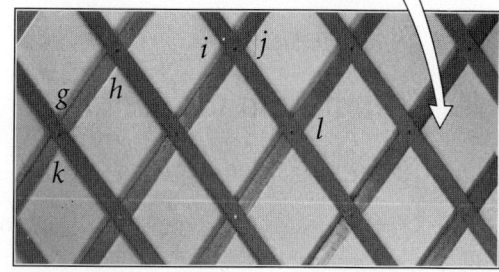

Contents

TG

Vertically opposite angles

Vertically opposite angles are the equal angles you get

when two straight lines cross, like this 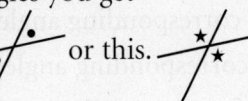 or this.

A2 (a) Which angle is vertically opposite to angle *g*?

(b) Which angle is vertically opposite to angle *b*?

(c) Give the letters for one more pair of vertically opposite angles.

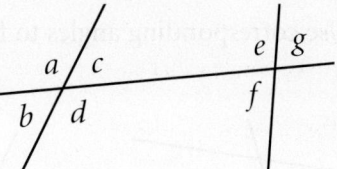

A3 Find four pairs of vertically opposite angles in the trellis pictures opposite.
Give their letters.

A4 Use vertically opposite angles to find the angles marked with letters here.

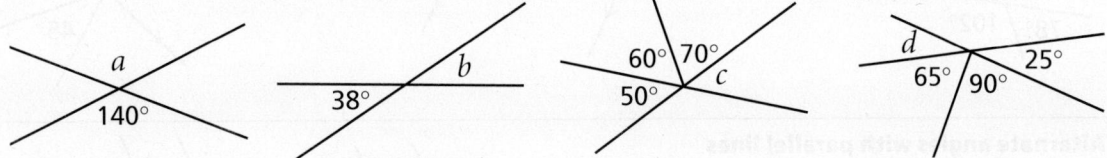

TG

Corresponding angles with parallel lines

The equal angles you get in a pattern like

this 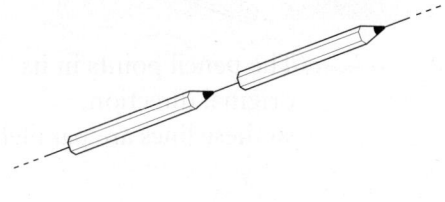 or this ...

are called corresponding angles.

The arrows show lines that are parallel.
The line that crosses the parallel lines is called a transversal.

To see how corresponding angles work,
think of two pencils in a straight line.

Now both pencils rotate 70° clockwise
about their ends.

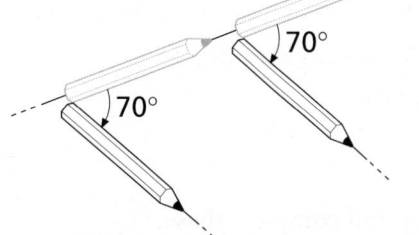

The pencils point in the same direction, so
these lines are parallel.

A5 Find four pairs of corresponding angles in the trellis pictures opposite.
Give their letters.

A6 Copy and complete these.

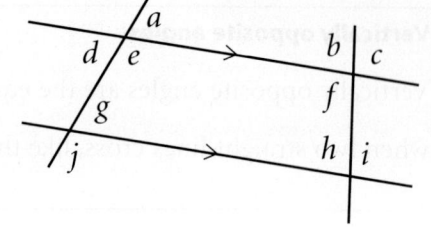

(a) Angles *a* and ___ are corresponding angles.

(b) Angles *h* and ___ are corresponding angles.

(c) Angles ___ and *i* are corresponding angles.

(d) Find one more pair of corresponding angles.

A7 Use corresponding angles to find the angles marked with letters here.

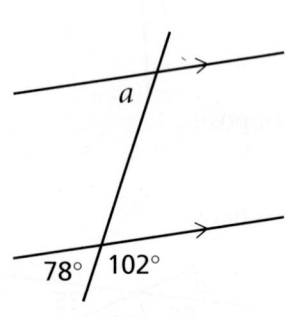
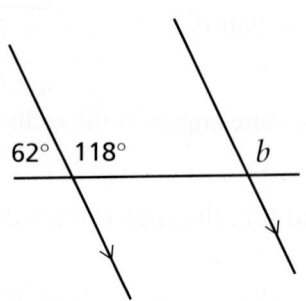
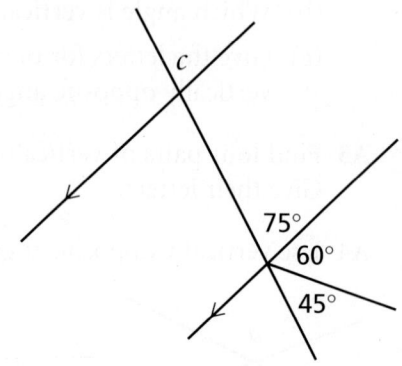

Alternate angles with parallel lines

Alternate angles are equal angles in a pattern like this 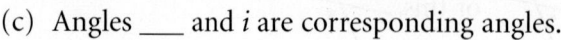 or this.

To understand alternate angles, think of one pencil.

It rotates 70° clockwise about its end ...

... then 70° anticlockwise about its point.

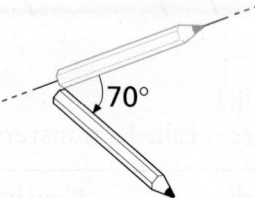

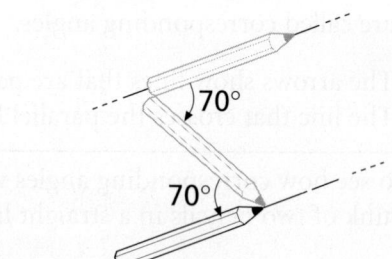

The pencil points in its original direction, so these lines are parallel.

A8 Copy and complete these.

(a) Angles ___ and *e* are alternate angles.

(b) Angles *b* and ___ are alternate angles.

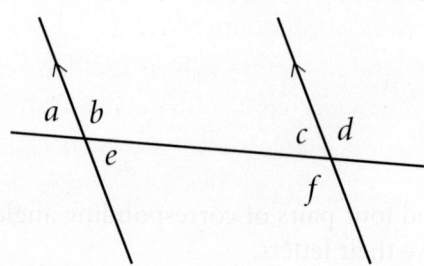

A9 Find four pairs of alternate angles in this diagram.
Give their letters.

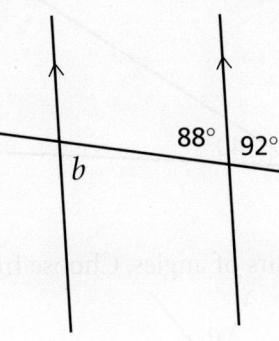

A10 Use alternate angles to find the angles
marked with letters here.

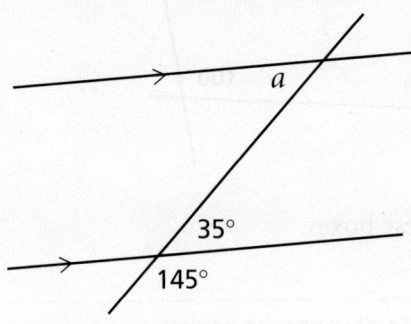

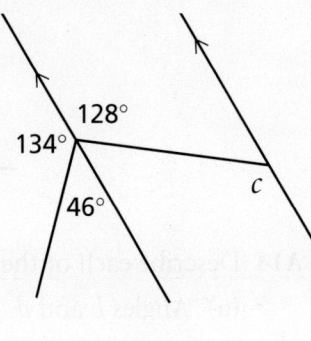

Supplementary angles

Supplementary angles add up to 180°.

You get supplementary angles on a straight line

or between parallel lines like this   or this.

A11 Copy and complete these.

(a) Angles ___ and *d* are supplementary angles
on a straight line.

(b) Angles *g* and ___ are supplementary angles
between parallel lines.

(c) Find one more pair of supplementary angles
on a straight line. Give their letters.

(d) Find one more pair of supplementary angles
between parallel lines. Give their letters.

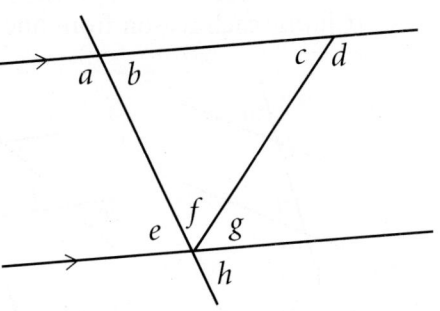

A12 Find three pairs of supplementary angles in this diagram.
Give their letters.

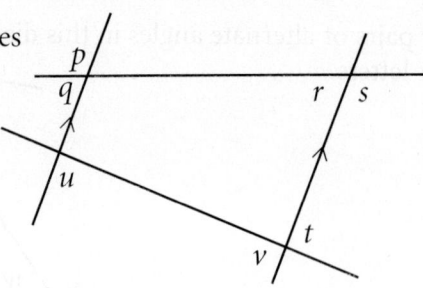

A13 Use supplementary angles to find the angles marked with letters here.

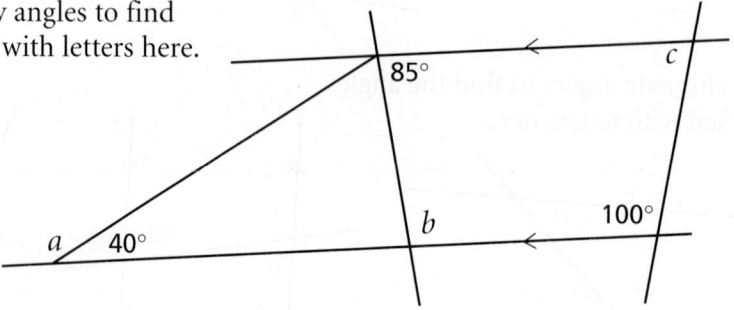

A14 Describe each of these pairs of angles. Choose from these boxes.

(a) Angles *b* and *d*
(b) Angles *b* and *f*
(c) Angles *c* and *g*
(d) Angles *a* and *b*
(e) Angles *e* and *f*
(f) Angles *c* and *e*
(g) Angles *e* and *g*
(h) Angles *a* and *e*
(i) Angles *d* and *g*

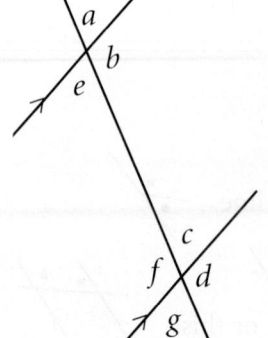

| Vertically opposite angles |
| Corresponding angles |
| Alternate angles |
| Supplementary angles on a straight line |
| Supplementary angles between parallel lines |

A15 Give the value of each lettered angle and the reason you know the angle.
(Choose each reason from one of the boxes in question A14.)

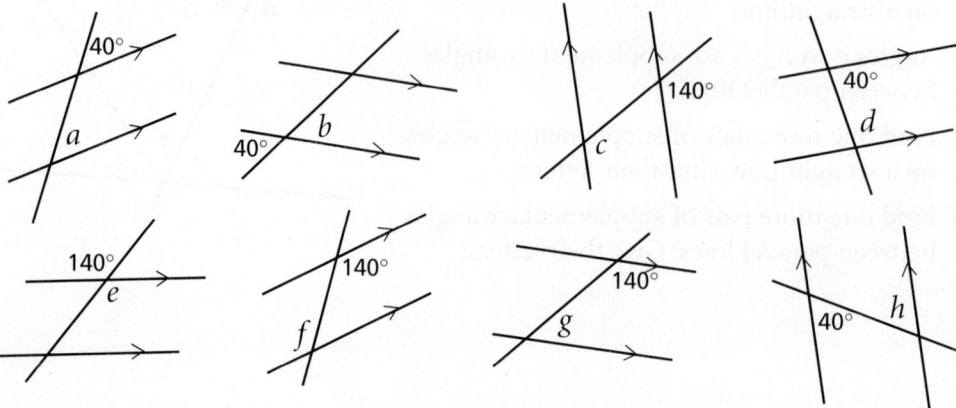

A16 Work out each angle marked with a ?.
Give your reason.

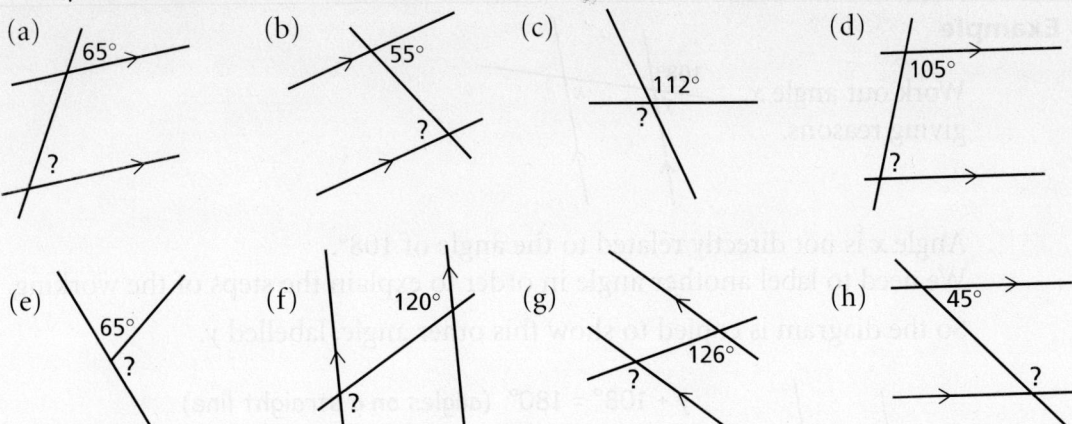

(a) 65° ?

(b) 55° ?

(c) 112° ?

(d) 105° ?

(e) 65° ?

(f) 120° ?

(g) 126° ?

(h) 45° ?

A17 Work out each lettered angle.
Give a reason.

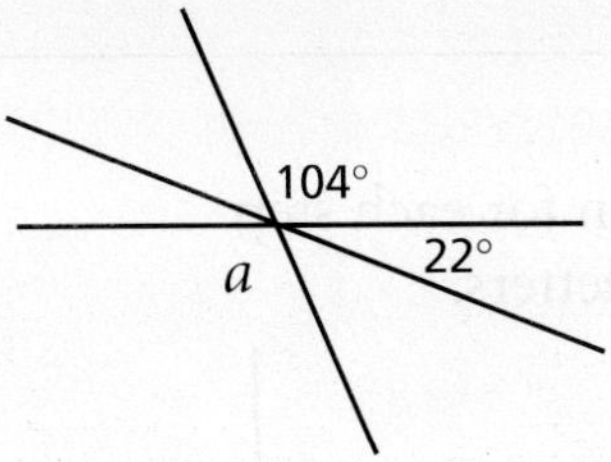

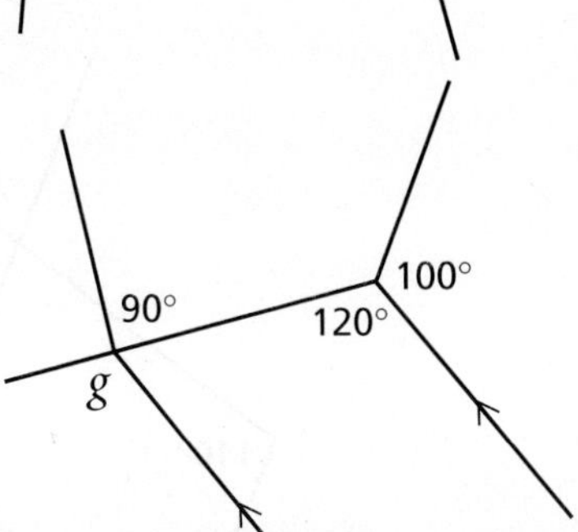

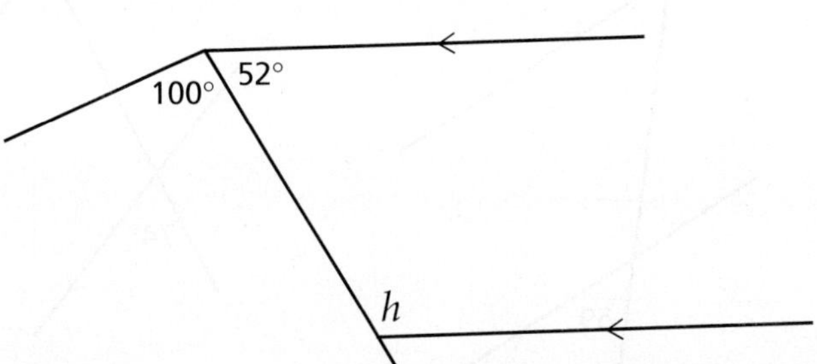

104° 22° *a*

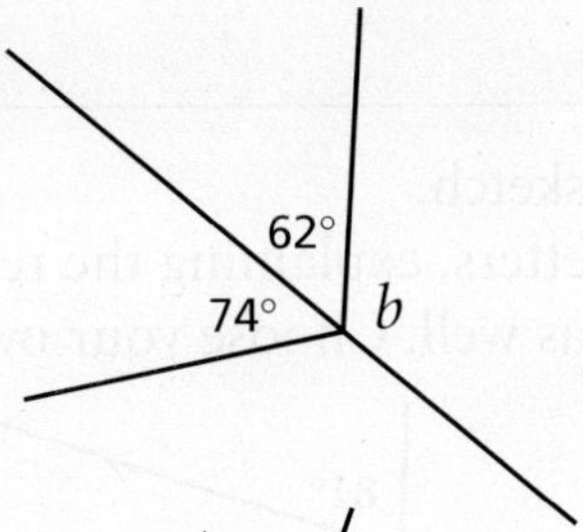

62° 74° *b*

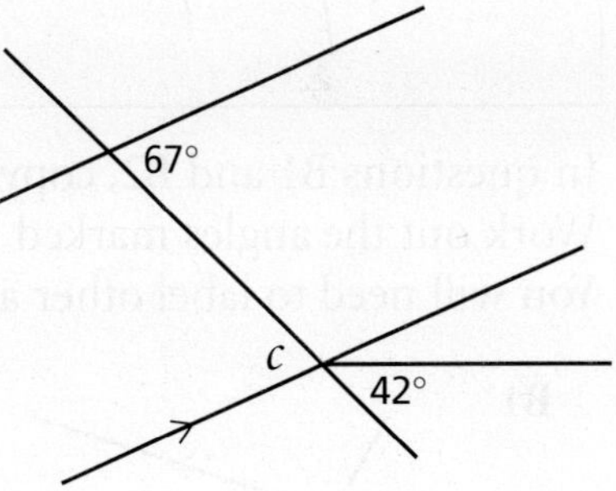

67° *c* 42°

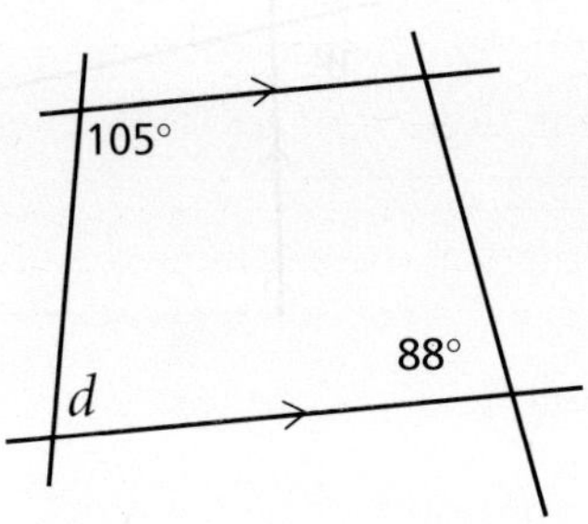

105° 88° *d*

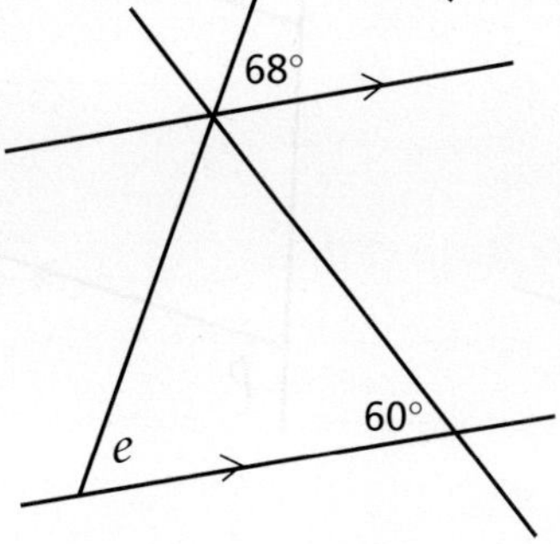

68° 60° *e*

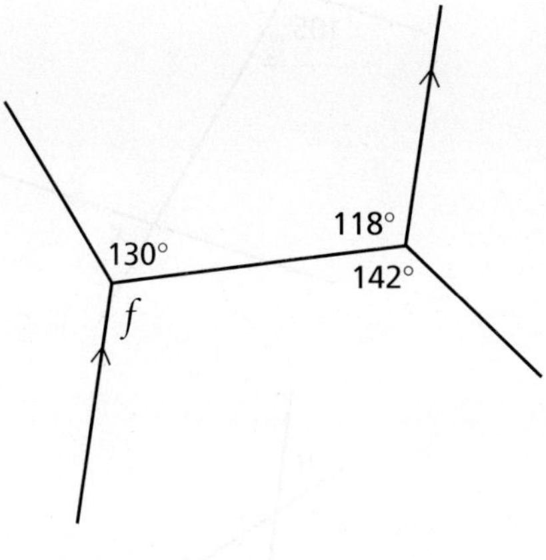

130° 118° 142° *f*

90° 100° 120° *g*

100° 52° *h*

B *Explaining how you work out angles*

Example

Work out angle *x*, giving reasons.

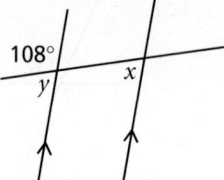

Angle *x* is not directly related to the angle of 108°.
We need to label another angle in order to explain the steps of the working.

So the diagram is copied to show this other angle, labelled *y*.

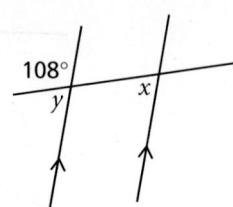

y + 108° = 180° (angles on a straight line)

So *y* = 72°

x = *y* (corresponding angles)

So *x* = **72°**

In questions B1 and B2, copy each sketch.
Work out the angles marked with letters, explaining the reason for each step.
You will need to label other angles as well. Choose your own letters.

B1

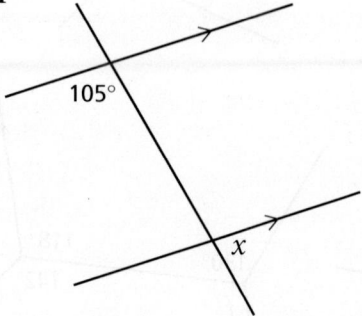

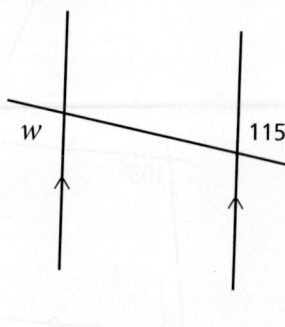

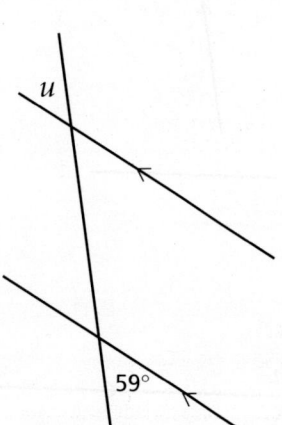

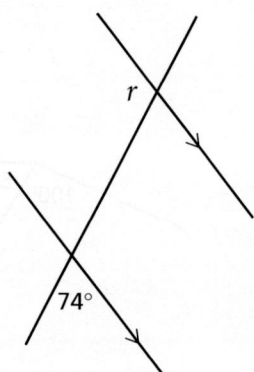

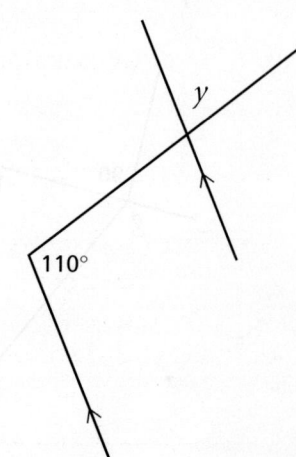

B2

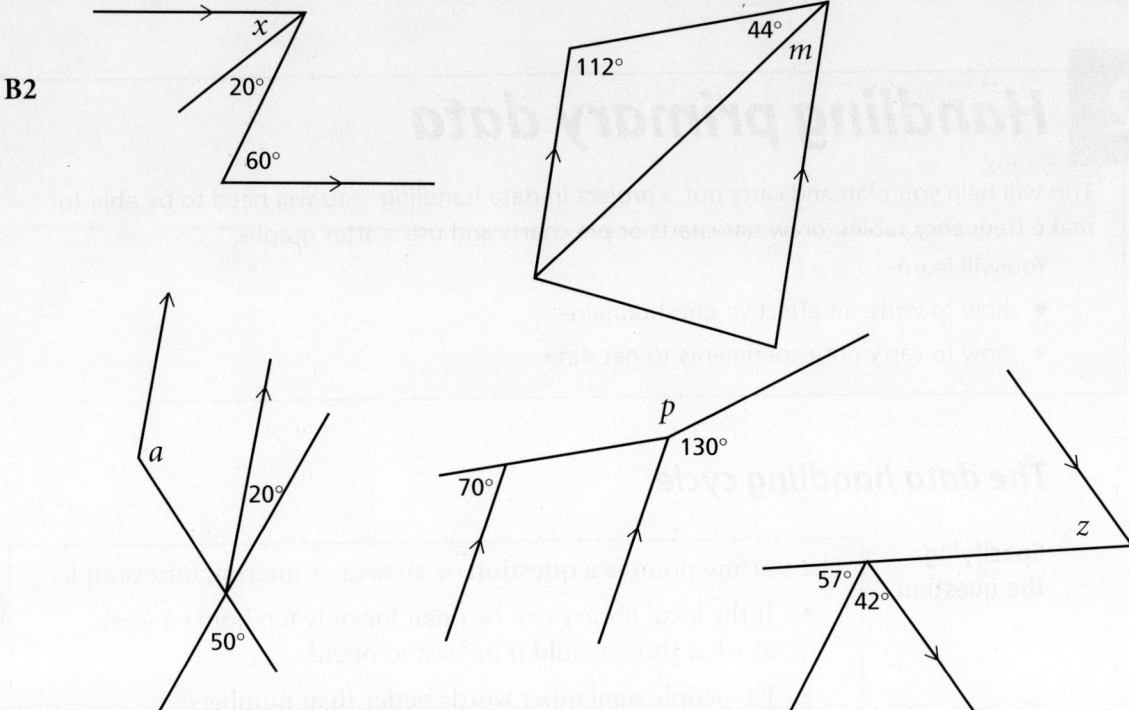

What you should have learned from this unit

- ◆ to identify vertically opposite, corresponding, alternate and supplementary angles
- ◆ to work out angles related to parallel lines
- ◆ to explain how you have worked out angles

Test yourself with these questions

T1 Find four pairs of alternate angles in this diagram. Give their letters.

T2 Give the values of the lettered angles. Give a reason.

T3 Copy this sketch.
Work out the angle marked *x*, explaining your reasons.
Label any other angles you need for your explanation.

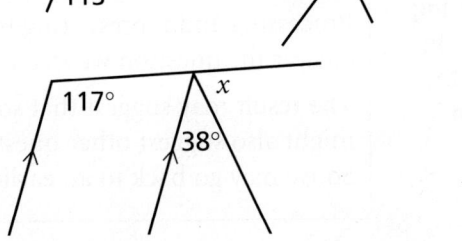

2 Handling primary data

This will help you plan and carry out a project in data handling. You will need to be able to make frequency tables, draw bar charts or pie charts and use scatter graphs.

You will learn

◆ how to write an effective questionnaire

◆ how to carry out experiments to get data

A The data handling cycle

Specifying the question	The starting point is a **question** or an area of interest, for example: • If the local library can be open for only ten hours a week, at what times would it be best to open? • Do people remember words better than numbers?
Collecting data	To answer the question, we need to decide what information or data we need. We have to plan how to **collect** it and how we will use it to help answer our question. If we have to collect the data ourselves, for example by asking people questions or by counting or measuring something, then the data is called **primary data**. If the data has already been collected by someone else, it is called **secondary data**.
Processing and representing the data	To help answer the question, the data has to be **processed**, (for example, by working out percentages, finding frequencies, calculating means, and so on). It is often helpful to **represent** the data in pictorial form (for example, frequency chart, scatter diagram, pie chart).
Interpreting the data to answer the question	Processing and representing the data allows us to **interpret** it to help answer the question we started with. The result may suggest that some more data needs to be collected. It might also suggest other questions which need answering. So we may go back to an earlier stage of the cycle and repeat.

Primary data and secondary data

Primary data is data which you collect yourself.
For example, you are collecting primary data when you give
people questionnaires to fill in. You are also collecting
primary data when you make measurements in an experiment.

Height	152 cm	172 cm
Weight	57 kg	78 kg
Pulse rate before exercise	89 bpm	75 bpm
Pulse rate after exercise	127 bpm	133 bpm

Secondary data is data which someone else has collected and organised.
For example, data about crime which is published by the Government is
secondary data.

Vandalism, per 10000 households			
Year	1981	1993	1995
Cases	1481	1638	1614

Sometimes data does not fit easily into either type.
For example, suppose you collect information about the prices of
secondhand cars from newspaper adverts. Is this primary or secondary data?
It feels more like primary data because although it's written by someone else,
it isn't organised in any way.

Ford Fiesta 1.4, 1992, 65000 miles. Blue vgc.
MOT. £1250 ono.
Ford Sierra 2.0LX, J reg, 1992, 34000 miles. Red.
One owner. fsh. MOT, taxed to Aug. £2600.
Ford Sierra 1.8 estate, 1994, Grey. No rust. recent
service. £2900 ono.

B Surveys

School uniform   Report by Chris and Melanie

The school council discussed changing the school uniform. Some people didn't like the colour and some wanted sweatshirts instead of blazers. A lot of pupils thought that there shouldn't be a uniform at all.

We decided to find out what other students felt about the uniform. We thought that boys and girls might feel differently and so might different year groups. We wrote a questionnaire and we decided to give it to some students in every year group. (There are about 180 in each year.)

Here is our questionnaire.

1 What year group are you in? (Please tick.)   Y7   Y8   Y9   Y10   Y11

2 Are you male or female?   Male   Female

3 Do you think there should be a school uniform?   Yes   No

4 If there has to be a uniform, would you prefer   blazer   sweatshirt

5 What colours would you like the uniform to be?

6 "Students should be allowed to wear jewellery." What do you think?

   Strongly agree   Agree   Not sure   Disagree   Strongly disagree

---

Questions for discussion

◆ What do you think of the questions? Are they easy to answer?
Are they clear - will they mean the same to everyone who answers them?
Will the responses be easy to analyse?

◆ Who would you give the questionnaire to?
How would you collect their responses?
How many people would you give it to?

I'll ask everyone in the school.

I'll ask all my friends.

I'll ask 5 people in each year.

I'll ask everyone in the choir.

The report continues like this:

Our teacher told us it was a good idea to pilot a questionnaire. This means giving it to a few people to see if there are any "bugs" (problems).

We gave it to 10 people. Some of them thought there should be a question about ties. Two people said that "jewellery" was too vague: ordinary rings could be allowed but not nose rings.

We also found that everybody had written different colours that they liked, sometimes three or four colours, eg dark blue, red, yellow. It would be difficult to analyse the answers to this question.

◆ Look back at the questionnaire.
  How could you improve it to avoid these problems?

In their report, Chris and Melanie made tables of the replies they got to the questions in their questionnaire.

This table shows the replies we got to the question "Would you prefer blazer or sweatshirt?"

Year		7	8	9	10	11
Boys	Blazer	7	7	4	3	3
	Sweatshirt	8	11	12	11	13
Girls	Blazer	7	7	6	6	7
	Sweatshirt	5	8	9	10	9

B1 Draw a chart, or charts, to illustrate this data. Explain why you chose your type of chart.

B2 What conclusions would you draw about the preferences?

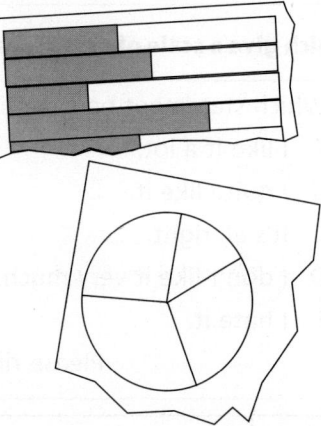

Question types

Here are some types of question you could use in a questionnaire.
Questions which ask for boxes to be ticked (or letters to be ringed)
make it easier to collect all the data together afterwards.

Yes/no questions

a	Have you passed the driving test?	YES ☐ NO ☐

The question must have a clear yes or no answer.
(If you think someone might not know the answer,
then you could include DON'T KNOW ☐).

Multiple-choice questions

b	Which age group are you in?
	0 - 19 ☐ 20 - 39 ☐ 40 - 59 ☐ 60 - 79 ☐ 80 or over ☐

c	Which of these statements best describes how you plan what you will watch on TV?
	A I plan days ahead.
	B I decide on the day.
	C I just flick around to see what's on.
	Please ring A B C

In examples *b* and *c* above, the person chooses one response.
In example *d* below, they can choose more than one.

d	Which of these languages do you study for GCSE?
	French ☐ German ☐ Spanish ☐ Gujerati ☐ Latin ☐

Questions which give a scale of responses

e	Which statement best descibes how you feel about maths?
	A I like it a lot.
	B I quite like it.
	C It's all right.
	D I don't like it very much.
	E I hate it.
	Please ring A B C D E

Questions which ask for a number

f	How many subjects are you taking at GCSE?	Number: ....

If you don't need to know the number exactly,
then it is better to give groups (as in example *b*).

Questions which ask for an order of preference

g	What kind of music do you prefer?
	Put in order of priority (1 for your favourite, 5 for your least).
	Hard rock ☐ Pop ☐ Jazz ☐ Easy listening ☐ Classical ☐

Open questions

h	What do you think about school lunches?

This kind of question is good for finding out people's own ideas.
But it is hard to summarise the answers.

Things to avoid!

◆ Don't ask questions which could be embarrassing. ('How old are you?')

◆ Don't ask questions which try to lead people to answer in one way.
  ('Would you like to see the safety of our children improved by banning traffic
  from the road in front of the school?') These are called **leading questions**.

◆ Don't ask questions which are difficult to answer precisely.
  ('How many hours of TV do you usually watch each week?')

B3 Criticise these questions and try to improve them.

How much do you earn?   £...............

How many are there in your family? ....

Where do you shop? Please tick.   Asda ☐   Sainsbury's ☐   Safeway ☐   Tesco ☐

How much do you spend a week on food?   £...............

How do you think supermarket fruit and vegetables compare with
the real fruit and vegetables you buy direct from a farm?

Carrying out a survey

1   Be clear about the purpose of your survey.

2   Write a draft questionnaire.

3   Pilot your draft questionnaire with a small number of people.

4   Improve the questions if necessary.

5   Decide who to give the questionnaire to, and how many people to ask.

6   Decide whether you will see people and ask the questions, or give them the questionnaire to fill in.

7   Collect all the responses together, analyse them and write a **report**.

In your report:

◆ State the purpose of your survey. Describe how you carried it out, any difficulties you had to overcome and any changes of plan.

◆ Include your final questionnaire.

◆ Say how many people responded.

◆ Summarise the responses to each question. Use tables and charts where appropriate.

If you are comparing the responses of different groups (eg boys and girls), summarise them **separately**. You could use a table something like this:

Hours of TV	0-9	10-19	20-29	30+
Girls	17	12	15	9
Boys	12	10	19	10

◆ Write a conclusion.

Points for discussion

Music charts

January 1965

1 I FEEL FINE — Beatles
2 DOWNTOWN — Petula Clark
3 WALK TALL — Val Doonican
4 SOMEWHERE — P.J.Proby
5 I'M GONNA BE STRONG — Gene Pitney
6 YEH YEH — Georgie ... plus Flames
7 I COULD EA... WITH YOU...
8 NO ARMS ...
9 TERRY
10 I UNDERS...

January 1975

1 DOWN DOWN — Status Quo
2 NEVER CAN SAY GOODBYE — Gloria Gaynor
3 STREETS OF LONDON — Ralph McTell
4 THE BUMP — Kenny
5 MS GRACE — Tymes
6 I CAN HELP — Billy Swan
7 ARE YOU READY TO ROCK

January 1985

1 DO THEY KNOW IT'S CHRISTMAS — Band Aid
2 EVERYTHING SHE WANTS — Wham!
3 NELLIE THE ELEPHANT — Toy Dolls
4 LIKE A VIRGIN — Madonna
5 WE ALL STAND TOGETHER — Paul McCartney & the Frog Chorus
6 EVERYTHING MUST CHANGE — Paul Young
   ...OVE ...es To Hollywood
   ...Tears Fo...

> The first British music chart was published in 1952.
>
> In the early days the record charts were based on the number of records sold in only a few hundred shops. Nowadays there are over 3000 record stores involved. This is about 75% of the record stores in Britain.

Why so many stores?

US Elections

Surveys got a bad name in 1936. In that year the U.S. Presidential elections were held. There were two candidates, Landon (who represented the better off) and Roosevelt (for the less well off).
A magazine did a postal survey on who people would vote for. They obtained the names and addresses from telephone directories and car registrations.
Over 2 million of the 10 million sent questionnaires replied. These predicted a massive victory for Landon.
In fact Roosevelt won by a massive majority!

Why do you think the result was so different?

Honest!

About 60 years ago, an American survey contained the question:

What do you think of of the new metallic Metals Law?

The option boxes included 'I don't know' as an option, but less than 25% ticked it. Everyone else ticked an opinion.

In fact the 'new metallic Metals Law' was completely fictitious!

Why did over 75% of people express an opinion?

C Experiments

Priya and Ben decide to investigate how good people are at remembering words, numbers and pictures.
They write a report on their findings.

Remembering words, pictures and numbers
by Priya and Ben

We wanted to see if there was any difference between how good young people were at remembering words, pictures and numbers.

We both thought it would be easiest to remember pictures.

We decided to test Years 10 to 13 who are mostly between 14 and 18 years old.

How we got our results
We made up some experiments.

We chose:
- 10 words - we tried to make sure there were no links between them (like 'pencil' and 'paper')
- 10 pictures
- 10 numbers between 1 and 100

We showed our class the 10 words for 30 seconds and gave them 60 seconds to write down as many as they could remember. The order didn't matter.

We did the same with the pictures and the numbers.

Each correct word, picture or number scored 1 point.

Each student had three scores out of 10 and wrote them on a slip of paper. Our class is in Year 10 and we wanted results from Years 10 to 13. We couldn't use Year 11 because they were on exam leave so we asked our teacher Mr Cassell to do the same experiment on his Year 12 and 13 mathematics groups.

Our results
We collected all the slips of paper and chose 10 at random from each year so that we had the results for 30 students.

school
heather
lamp
sky
hate
spoon
necklace
birthday
hair
leaf

Here are some questions for discussion.
Explain each of your answers as fully as you can.

C1 Did you find Priya and Ben's description of their memory experiments easy to follow ?

C2 They made up a list of 10 words for one experiment.
Why do you think they tried to make sure there were no links between their words ?

C3 Why do you think they used the same number of words, numbers
and pictures in their experiments ?

C4 Was it a good idea for Mr Cassell to collect the data from his
Year 12 and 13 mathematics groups ?

C5 Why do you think they chose 10 students at random from each year ?
Do you think this was a good idea ?

We made a table for each year but we analysed all the results together.
W stands for Words; P stands for Pictures; N stands for Numbers

Year 10		
W	P	N
8	7	5
5	6	6
5	3	7
7	7	5
8	7	7
10	10	8
9	10	8
10	10	9
10	9	10
9	8	6

Year 12		
W	P	N
8	9	6
9	10	9
10	10	8
9	9	9
7	9	8
10	10	7
7	8	4
7	8	10
8	7	8
9	10	7

Year 13		
W	P	N
10	**10**	**9**
9	9	6
9	10	10
7	9	4
8	10	7
8	9	5
8	10	4
8	8	4
8	8	6
10	10	6

Each row shows the scores for one student.
For example, the first row in the Year 13 table shows that a student correctly remembered 10 words, 10 pictures and 9 numbers.

Analysing our results

We drew bar charts for our results.

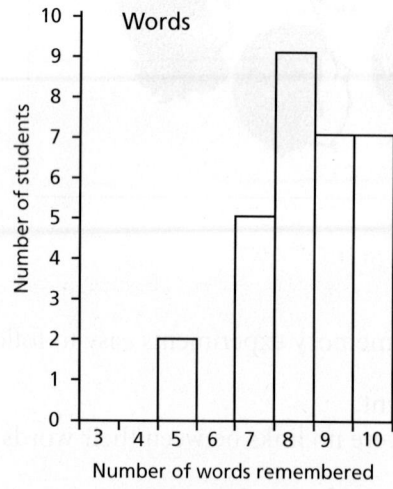

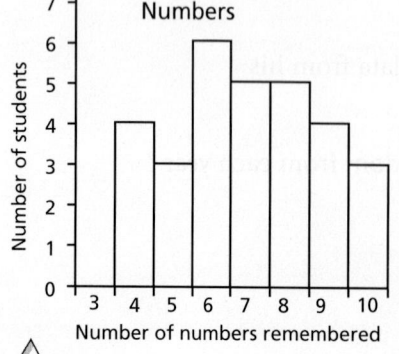

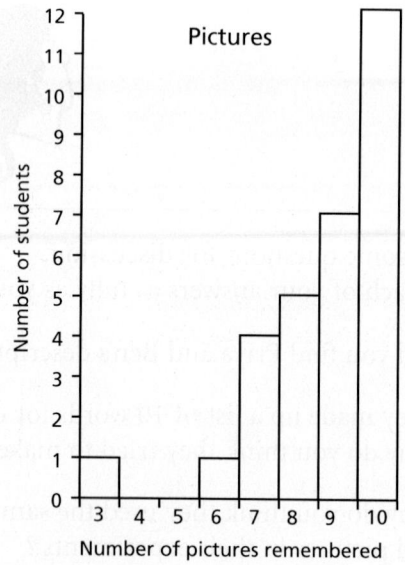

Conclusion

From the shape of our charts, we think that young people are best at remembering pictures, then words and then numbers. We expected that people would be best at remembering pictures.

For this sample:

C6 One student remembered only 3 pictures.
How many words and numbers did this person remember?

C7 How many students in Year 13 remembered all 10 pictures?

C8 How many students from all three years remembered less than 7 pictures?

C9 How many students from all three years remembered more than 8 words?

C10 Do you think that the bar charts show that the students are best at remembering pictures, then words and then numbers ?

We then decided to calculate the mean, median and range for each set of results.

	Words	Pictures	Numbers
Mean	250 ÷ 30 ≈ **8.3** words	260 ÷ 30 ≈ **8.7** pictures	208 ÷ 30 ≈ **6.9** numbers
Median	**8** words	**9** pictures	**7** numbers
Range	10 – 5 = **5** words	10 – 3 = **7** pictures	10 – 4 = **6** numbers

The means and medians show that our conclusion is correct.
Young people are best at remembering pictures, then words and then numbers.
The ranges show that the results for the pictures are more spread out.

C11 Why do they think that the values for the means and medians show that their conclusion is correct ?

C12 Do you think they have enough evidence to say that young people are best at remembering pictures, then words and then numbers ?

C13 Why do you think they did not compare students from Years 10, 12 and 13 ?

Extension

We then investigated if there was a link between
our memory for words and for numbers.

We drew a scatter diagram for our results.

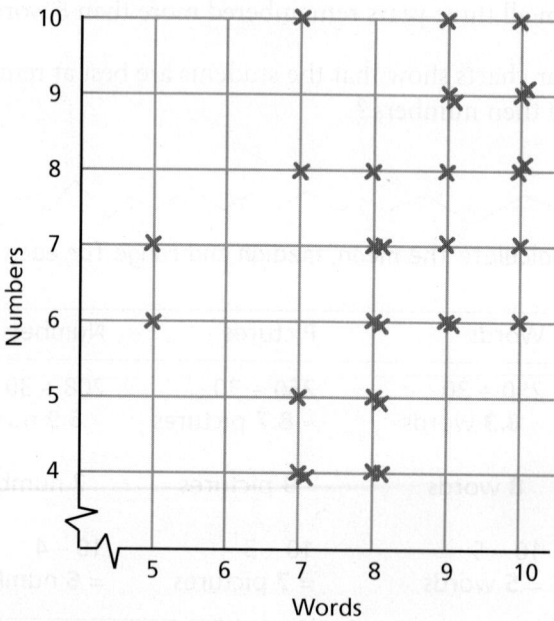

Conclusion

The crosses are quite spread out.
We don't think this shows a link between our memory for words and for numbers.

Possible further work

If we had more time we would have tried to find if there was a link between
our memory for words and for pictures and also for numbers and pictures.

We also think that we would like to do the experiments again with more than
10 words, pictures and numbers. A lot of people scored 10 in at least one
experiment.

Next week we are going to ask our class how many of the words, pictures and
numbers they can remember (but we haven't told them this). We want to see
how good they are at remembering these things after a week has passed.

C14 Do you agree with Priya and Ben that the scatter diagram shows there is no link between our memory for words and for numbers?

C15 For the students in Priya and Ben's investigation, draw a scatter diagram to decide if there is a link between their memory for words and for pictures.

C16 Why do you think Priya and Ben want to repeat their experiments with more than 10 numbers, words and pictures?

C17 Investigate to see if you come to the same conclusions as Priya and Ben for students in your school.

D *Ideas for projects*

These projects are described in more detail on sheets P01 to P09

Remember, remember …

Investigate aspects of memory, such as whether age affects memory or whether background music makes it easier to remember.

Lunchtime menu

Carry out a surey of people's eating habits in order to decide what to include in the lunchtime menu of a cafe.

Food for thought

Investigate aspects of healthy and unhealthy eating.

First names

Investigate aspects of people's first names, such as popularity or length.

Groovers

Investigate people's preferences for different styles of music, for example to programme the output of a radio station.

Computer games

Investigate opinions about computer games and their popularity.

Wine gums

Compare different makes of wine gum: cost, taste, and so on.

Town and country houses

Use the information given in estate agents' adverts to investigate aspects of houses, such as how their value varies from place to place.

Helicopter seeds

Some trees have seeds with wings that rotate in the wind as the seed falls. Investigate how these seeds fly by making simple paper models.

3 Working with expressions

You should know

◆ how to evaluate expressions such as $2 - 7 + 3$ and $7 - 2 \times 3$

◆ that $6n$ means $6 \times n$ and $\dfrac{12a}{4}$ means $12a \div 4$

You will

◆ Substitute into simple linear expressions

◆ Simplify expressions such as $4n \times 6$ and $\dfrac{8n}{2}$

◆ Simplify expressions such as $2x + 1 + 3x - 7$

◆ Multiply out brackets in expressions such as $4(6n + 1)$

◆ Simplify expressions such as $\dfrac{8n - 4}{2}$

◆ Use algebra to prove general statements like 'The result for this puzzle will always be 2'

A Substitution

- Evaluate any expressions in brackets first.
- Then work out any multiplications and divisions.
- Then work out any additions and subtractions.

Examples

Find the value of
$2x + 5$ when $x = 3$.

$$
\begin{aligned}
2x + 5 &= 2 \times 3 + 5 \\
&= 6 + 5 \\
&= 11
\end{aligned}
$$

Find the value of
$3(x - 1)$ when $x = 5$.

$$
\begin{aligned}
3(x - 1) &= 3 \times (5 - 1) \\
&= 3 \times 4 \\
&= 12
\end{aligned}
$$

Find the value of
$\dfrac{14 - 3h}{2}$ when $h = 2$.

$$
\begin{aligned}
\frac{14 - 3h}{2} &= \frac{14 - 3 \times 2}{2} \\
&= \frac{14 - 6}{2} \\
&= \frac{8}{2} = 4
\end{aligned}
$$

A1

p $2(n + 1)$ q $3n + 1$ r $12 - n$ s $3(5 - n)$ t $10 - \dfrac{n}{2}$

u $\dfrac{n + 1}{2}$ v $\dfrac{n}{2} + 1$ w $10 - 2n$

(a) Find the value of each expression when $n = 4$.

(b) Which expressions have a value of 0 when $n = 5$?

(c) Which expression has the greatest value when $n = 3$?

(d) Which expression has the lowest value when $n = 1$?

A2 Find the value of the following expressions when $a = 10$

(a) $2a - 3$ (b) $15 - a$ (c) $30 - 2a$ (d) $2(a + 7)$

(e) $3(2a - 5)$ (f) $\dfrac{2a - 5}{3}$ (g) $\dfrac{2a + 4}{6}$ (h) $10 - \dfrac{3a}{5}$

A3 Each expression in the diagram stands for the length of a side in centimetres.

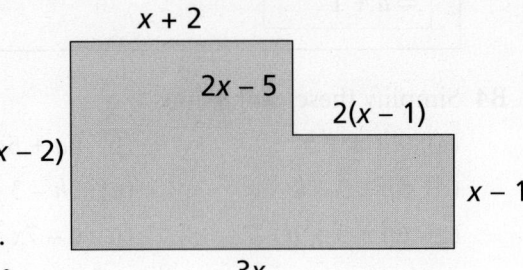

(a) What is the length of the longest side when $x = 3$?

(b) (i) Work out the length of each side when $x = 3$ and draw the shape.

    (ii) What is the perimeter of your shape?

(c) What is the perimeter of the shape when $x = 5$?

B Simplifying

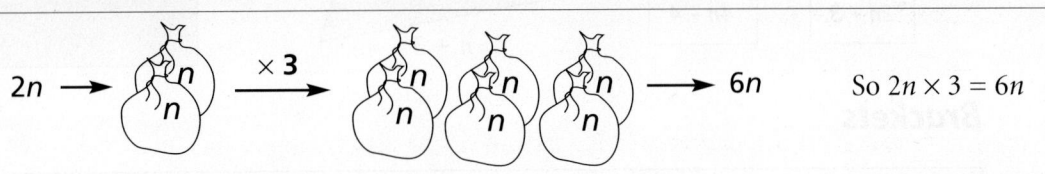

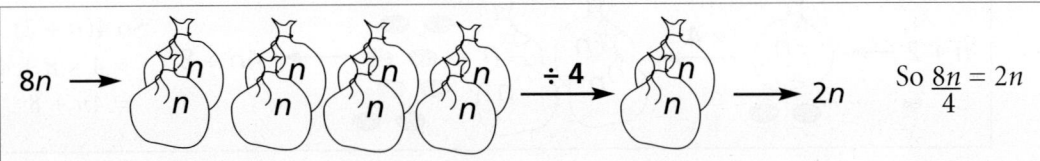

B1 Simplify these expressions.

(a) $2 \times 5n$ (b) $3 \times 2y$ (c) $6a \times 5$ (d) $4 \times 7b$ (e) $4x \times 9$

(f) $\dfrac{4n}{2}$ (g) $\dfrac{6a}{3}$ (h) $\dfrac{15y}{5}$ (i) $\dfrac{20x}{4}$ (j) $\dfrac{36b}{9}$

B2 Which of these is an expression for the area of this rectangle?

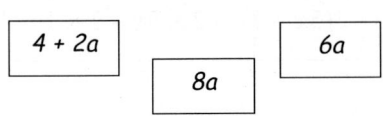

B3 Write an expression for the area of each of these rectangles.

Simplifying by collecting like terms
Examples

$6 + a - 5$	$7b - 3b + b$	$7 - 2p - 3 + 2p$	$3n - 5 - n - 6$
$= a + 6 - 5$	$= 5b$	$= 7 - 3 - 2p + 2p$	$= 3n - n - 5 - 6$
$= a + 1$		$= 4$	$= 2n - 11$

B4 Simplify these expressions.

(a) $2 + p + 5$ (b) $6q + 5q - 3q$ (c) $3 + w - 1 + w$

(d) $3k + 5 - k$ (e) $2h - 3 - 5$ (f) $3m - 5 - m + 5$

(g) $6n + 3 + n - 7$ (h) $6 - 7x + 7x$ (i) $5y + 1 - y - 4y$

B5 Which of these expressions give the perimeter of

(a) the triangle (b) the rectangle

$\boxed{4n + 2}$ $\boxed{4n + 6}$

$\boxed{2n + 3}$ $\boxed{4n - 4}$

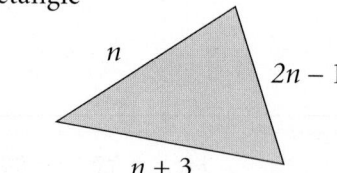

C *Brackets*

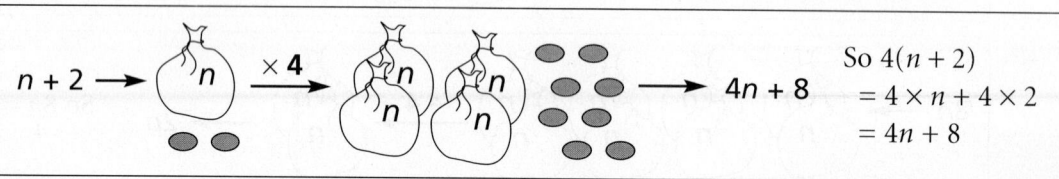

So $4(n + 2)$
$= 4 \times n + 4 \times 2$
$= 4n + 8$

$4(n + 2)$ and $4n + 8$ are **equivalent expressions**.

For example when $n = 3$

$4(n + 2)$	$4n + 8$
$= 4(3 + 2)$	$= 4 \times 3 + 8$
$= 4 \times 5$	$= 12 + 8$

$\searrow \quad 20 \quad \swarrow$

This works for any value of n.

Examples

$8(p - 3) = 8 \times p - 8 \times 3 = 8p - 24$

$3(5 + 2n) = 3 \times 5 + 3 \times 2n = 15 + 6n$

$2(5x - 1) = 2 \times 5x - 2 \times 1 = 10x - 2$

C1 Find four pairs of equivalent expressions.

(1) $4(a + 8)$ (2) $2(2a + 1)$ (3) $4(a + 2)$ (4) $4a + 2$

(5) $4a + 8$ (6) $4(a + 4)$ (7) $2(2a + 8)$ (8) $4a + 32$

C2 Multiply out the brackets from

(a) $6(n + 1)$ (b) $5(m - 4)$ (c) $3(5 + k)$ (d) $5(2c + 1)$ (e) $4(3h - 2)$

(f) $2(5a + 3)$ (g) $6(2w - 3)$ (h) $5(3 - p)$ (i) $4(3 - 8x)$ (j) $7(3c + 4)$

C3 (a) Which of these expressions gives the area of rectangle A?

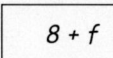 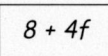

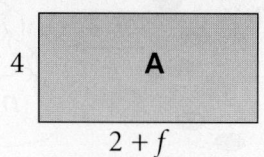

(b) Which of these expressions gives the area of rectangle B?

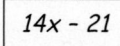

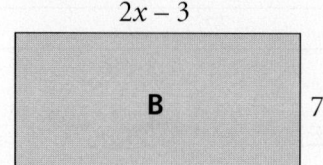

C4 For each shape, write expressions for the missing lengths.

(a)

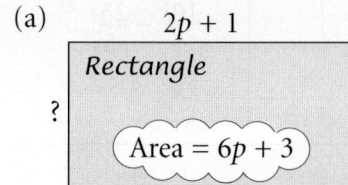

(b)

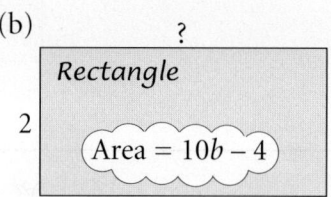

C5 Copy and complete:

(a) $3(\ldots + 5) = 6n + 15$ (b) $4(\ldots - p) = 8 - 4p$

(c) $5(\ldots - 3) = 20m - \ldots$ (d) $2(\ldots + \ldots) = 10x + 20$

C6 Ken and Fiona have £x each. They are each given £5.
Which expressions tell you the amount of money they have altogether?

| $5 + x$ | $2x + 5$ | $2(x + 10)$ | $2(x + 5)$ | $2x + 10$ |

C7 Jo has three orchards, each with n apple trees. Two trees in each field are blown down.
Find an expression for the total number of trees left.

C8 Sketch each shape and write expressions for the missing lengths.

(a)

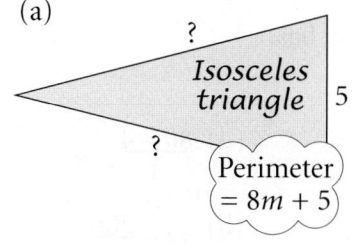

(b)

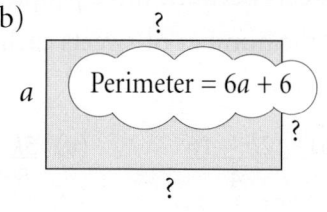

(c)

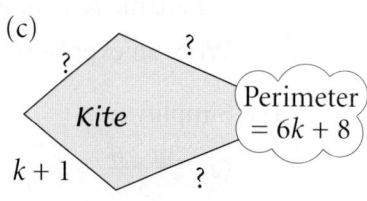

D Dividing

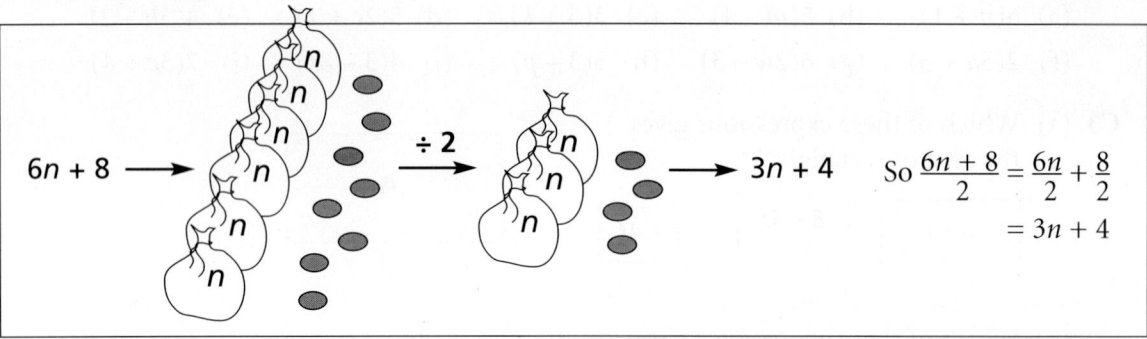

$$6n + 8 \longrightarrow \quad \div 2 \quad \longrightarrow 3n + 4 \qquad \text{So } \frac{6n + 8}{2} = \frac{6n}{2} + \frac{8}{2}$$
$$= 3n + 4$$

Examples

$$\frac{10n + 5}{5}$$
$$= \frac{10n}{5} + \frac{5}{5}$$
$$= 2n + 1$$

$$\frac{8x}{4} + 7$$
$$= 2x + 7$$

$$\frac{15n - 9}{3}$$
$$= \frac{15n}{3} - \frac{9}{3}$$
$$= 5n - 3$$

$$\tfrac{1}{5}(10k - 25)$$
$$= \frac{10k - 25}{5}$$
$$= \frac{10k}{5} - \frac{25}{5}$$
$$= 2k - 5$$

D1 I have four bags of sweets,
each with n sweets in them.
I also have 10 loose sweets.

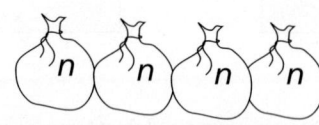 

I share these sweets between two people.
Write an expression for the number of sweets each person has.

D2 Simplify:

(a) $\dfrac{15n + 10}{5}$ (b) $\dfrac{7m}{7} + 3$ (c) $4 + \dfrac{18k}{6}$ (d) $\dfrac{12h}{3} + 2$

(e) $\dfrac{5p + 10}{5}$ (f) $\dfrac{4c + 6}{2}$ (g) $\dfrac{21y}{7} + 7$ (h) $\dfrac{18 + 12w}{3}$

D3 I have six bags of sweets, each with n sweets in them. I eat 18 sweets.
I share the remaining sweets between three people.

Write an expression for the number of sweets each person has.

D4 Simplify:

(a) $\dfrac{3a - 6}{3}$ (b) $\dfrac{12b - 16}{4}$ (c) $\dfrac{5k}{5} - 10$ (d) $\dfrac{8h - 4}{4}$

(e) $\dfrac{12d - 18}{6}$ (f) $\dfrac{20g - 30}{10}$ (g) $\dfrac{15 - 5m}{5}$ (h) $12 - \dfrac{20n}{4}$

D5 Copy and complete:

(a) $\dfrac{2m + \square}{2} = m + 7$ (b) $\dfrac{\square - 9}{3} = 2c - 3$ (c) $\dfrac{24 + 18y}{\square} = 4 + 3y$

D6 Copy and complete: $\frac{1}{3}(6n + 12) = \dfrac{6n + 12}{3} = 2n + ?$

D7 Simplify:

(a) $\frac{1}{2}(6n + 12)$ (b) $\frac{1}{3}(9x - 6)$ (c) $\frac{1}{4}(8k + 20)$ (d) $\frac{1}{5}(5p - 10)$

D8 Solve the puzzle on sheet P11.

*D9** Find an expression for the area of each triangle.

(a)

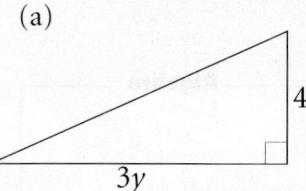

(b)

(c)

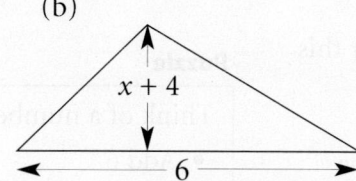

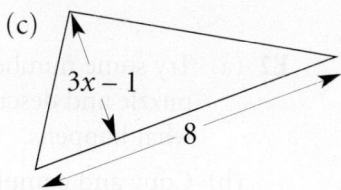

*D10** Find an expression for the **perimeter** of each rectangle.

(a)

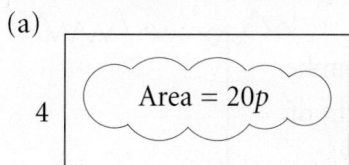

(b) $3a + 7$

(c)

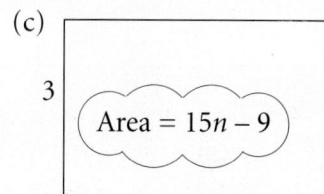

⊟ *Tricky business*

Think of a number
- Add 5
- Multiply by 2
- Subtract 7
- Subtract the number you first thought of
- Subtract 3

What is your final result?

Think of a number
- Add 4
- Multiply by 3
- Subtract 9
- Divide by 3
- Add 3
- Subtract the number you first thought of

What is your final result?

Think of a number
- Multiply by 4
- Subtract 2
- Multiply by 3
- Divide by 6
- Add 1
- Divide by 2

What is your final result?

- What happens with each puzzle?
- Can you use algebra to explain this?

TG

E1 (a) Try some numbers for this puzzle and describe what happens.

(b) Copy and complete the algebra box to explain how the puzzle works.

Puzzle

Think of a number
- Multiply by 6
- Add 3
- Divide by 3
- Subtract 1
- Divide by 2

What is the result?

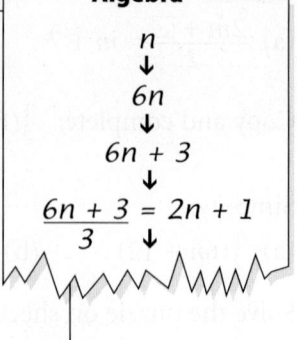

Algebra

$$n$$
$$\downarrow$$
$$6n$$
$$\downarrow$$
$$6n + 3$$
$$\downarrow$$
$$\frac{6n + 3}{3} = 2n + 1$$
$$\downarrow$$

E2 (a) Try some numbers for this puzzle and describe what happens.

(b) Copy and complete the algebra box to explain how the puzzle works.

Puzzle

Think of a number
- Add 6
- Multiply by 3
- Subtract 12
- Divide by 3
- Subtract the number you first thought of

What is the result?

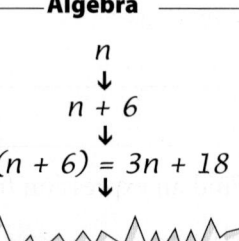

Algebra

$$n$$
$$\downarrow$$
$$n + 6$$
$$\downarrow$$
$$3(n + 6) = 3n + 18$$
$$\downarrow$$

E3 For each puzzle below

(a) Try some numbers and describe what happens.

(b) Use algebra to explain how the puzzle works.

1 Think of a number
- Add 1
- Multiply by 3
- Subtract 9
- Divide by 3
- Add 2

What is the result?

2 Think of a number
- Subtract 2
- Multiply by 4
- Add 8
- Divide by 4
- Subtract the number you first thought of

What is the result?

3 Think of a number
- Multiply by 6
- Add 15
- Subtract 3
- Divide by 3
- Add 6
- Divide by 2
- Subtract the number you first thought of

What is the result?

*E4 Try to make up your own puzzles like these.

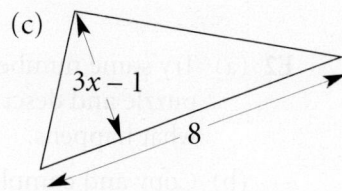

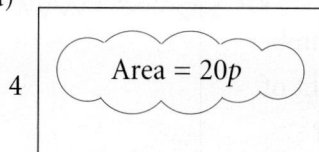
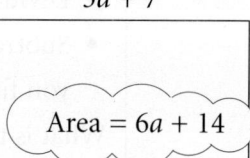
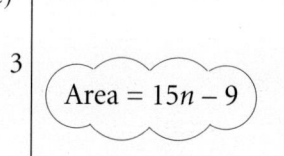

Test yourself with these questions

T1 Find the value of the following expressions when $n = 6$.

(a) $4n - 5$ (b) $5(n - 2)$ (c) $3(2n + 1)$ (d) $\dfrac{2n + 3}{5}$ (e) $10 - \dfrac{n}{2}$

T2 Multiply out the brackets from these expressions.

(a) $3(b + 6)$ (b) $2(5 - h)$ (c) $5(2a - 3)$ (d) $4(3x + 10)$

T3 Hal and Dwayne have n sweets each. They are each given 3 sweets.
Which expressions give the number of sweets they have altogether?

$3 + n$	$2n + 3$	$2(n + 3)$	$2(n + 6)$	$2n + 6$

T4 Find an expression for the
missing length in the rectangle.

?

4 Area $= 12p + 28$

T5 I have six bags of marbles, each with x marbles in them. I also have 12 loose marbles.
I share these marbles between three people.

Write an expression for the number of marbles each person has.

T6 Simplify the following expressions.

(a) $\dfrac{16n}{8}$ (b) $\dfrac{12m + 20}{4}$ (c) $\frac{1}{2}(8k - 10)$ (d) $\dfrac{14m}{7} - 7$

T7 (a) Try some numbers for
this puzzle and describe
what happens.

(b) Use algebra to explain
how the puzzle works.

> Think of a number
> - Subtract 3
> - Multiply by 6
> - Add 6
> - Divide by 6
> - Add 2
>
> What is the result?

T8 The perimeter of a regular hexagon is $12x - 18$.
What is an expression for the length of one edge?

T9 An expression for the length of one edge of a regular octagon is $3x + 5$.
What is the perimeter of this octagon?

Mental and written calculation

You should already know

◆ how to multiply or divide a number by 10, 100, 100, ... by moving the digits to the left or right (for example, $35.6 \div 100 = 0.356$)

◆ how to do multiplications like 26×34 and divisions like $476 \div 28$ without a calculator.

You will learn

◆ how to multiply and divide decimals without using a calculator.

Ⓐ *Multiplication*

Starting with **2 × 4 = 8**, we can get other multiplications.

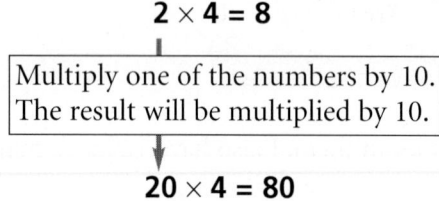

2 × 4 = 8

| Multiply one of the numbers by 10. |
| The result will be multiplied by 10. |

20 × 4 = 80

2 × 4 = 8

| Multiply both of the numbers by 10. |
| The result will be multiplied by 100. |

20 × 40 = 800

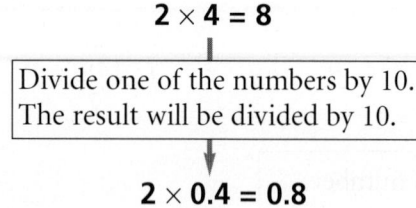

2 × 4 = 8

| Divide one of the numbers by 10. |
| The result will be divided by 10. |

2 × 0.4 = 0.8

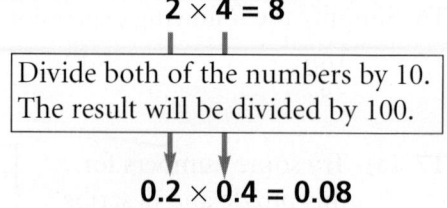

2 × 4 = 8

| Divide both of the numbers by 10. |
| The result will be divided by 100. |

0.2 × 0.4 = 0.08

Here is a chain leading from **3 × 5** to **0.3 × 500** :

3 × 5 = 15

3 × 50 = 150

3 × 500 = 1500

0.3 × 500 = 150

A1 Work these out.

(a) 3×0.4 (b) 0.3×0.4 (c) 30×0.4 (d) 3×0.04 (e) 30×0.04

(f) 30×0.3 (g) 3×0.03 (h) 0.3×0.3 (i) 300×0.3 (j) 0.03×3

A2 Work these out.

(a) 60×5 (b) 60×0.5 (c) 0.6×50 (d) 0.06×5 (e) 600×0.5

(f) 0.6×0.5 (g) 60×0.05 (h) 0.06×0.5 (i) 600×50 (j) 0.06×50

A3 These six 'cards' can be arranged to make two correct multiplications.
Show how to do it.

| 0.6 | = 24 | 6 | × 40 | × 0.4 | = 2.4 |

A4 Show how to arrange these nine cards to make three correct multiplications.

| 50 | × 400 | = 200 | × 50 | 0.4 | = 2000 | 0.5 | × 40 | = 20 |

A5 Arrange these cards to make four correct multiplications.

| = 6 | × 4 |

| 30 | × 0.3 | × 0.2 | = 0.06 | × 30 | = 1.2 | = 12 | 0.3 | 0.2 | 0.4 |

A6 The number 50 is fed into this 'chain' of number machines.

Input (50) —× 0.4→ ◯ —+ 20→ ◯ —− 10→ ◯ —× 0.2→ ◯ Output

(a) What is the output?

(b) What is the output if the machines are arranged in this order:

Input (50) —+ 20→ ◯ —× 0.2→ ◯ —− 10→ ◯ —× 0.4→ ◯ Output

(c) What is the largest output you can get by changing the order of the machines?
(The starting number is still 50.)

A7 The number 40 is fed into this 'chain' of machines.

Input (40) —× 0.5→ ◯ —+ 40→ ◯ —× 0.1→ ◯ —− 20→ ◯ Output

What is the largest output you can get by changing the order of the machines?

A8 Here is a set of numbers.

> 0.05   0.5   5   50   500
> 0.06   0.6   6   60   600

From this set, find as many pairs as possible whose product is:

(a) 30   (b) 3   (c) 300   (d) 25   (e) 2.5   (f) 0.36

A9 Do the calculation in each box.
Arrange the answers in order of
size, smallest first. The letters
will spell a word.

N	H	F	S
0.3×0.3	0.02×30	0.2×0.2	3×0.1

I	D	I	E
0.3×0.4	200×0.04	0.2×0.4	0.1×8

B *Written multiplication*

Example 3.4×0.26

| First get a rough idea of the answer, by rounding:

$\mathbf{3 \times 0.3 = 0.9}$ | Ignore the decimal points. Multiply 34 by 26.

Here is one way to do it.

$\begin{array}{r} 34 \\ \times\ 26 \\ \hline 680 \\ 204 \\ \hline 884 \end{array}$ | Decide where to put the decimal point in the result:

$\mathbf{3.4 \times 0.26 = 0.884}$ |

B1 Work these out.

(a) 42×3.8 (b) 1.7×2.7 (c) 0.58×26 (d) 0.13×5.6

(e) 9.2×0.16 (f) 0.45×0.88 (g) 2.6×0.014 (h) 0.83×0.43

B2 You are told that $156 \times 48 = 7488$.
Write down the answer to each of these.

(a) 1.56×48 (b) 15.6×4.8 (c) 0.156×48 (d) 1.56×0.48

B3 You are told that $127 \times 204 = 25908$.
Write down the answer to each of these.

(a) 12.7×20.4 (b) 1.27×2.04 (c) 0.127×204 (d) 12.7×0.204

C *Dividing by a decimal*

Sometimes you may need to divide by a decimal.

You can often change it to a whole number by multiplying 'top and bottom' by 10 or 100.

Examples

$$\frac{8}{0.2} \xrightarrow{\times 10}_{\times 10} = \frac{80}{2} = 40 \qquad\qquad \frac{1.5}{0.03} \xrightarrow{\times 100}_{\times 100} = \frac{150}{3} = 30$$

Work these out.

C1 (a) $\dfrac{6}{0.3}$ (b) $\dfrac{12}{0.2}$ (c) $\dfrac{18}{0.6}$ (d) $\dfrac{30}{0.5}$ (e) $\dfrac{2.4}{0.3}$

    (f) $\dfrac{1.4}{0.2}$ (g) $\dfrac{3.2}{0.8}$ (h) $\dfrac{0.18}{0.3}$ (i) $\dfrac{0.06}{0.3}$ (j) $\dfrac{120}{0.4}$

C2 (a) $\dfrac{1.2}{0.03}$ (b) $\dfrac{2.8}{0.04}$ (c) $\dfrac{16}{0.08}$ (d) $\dfrac{0.8}{0.02}$ (e) $\dfrac{8}{0.04}$

    (f) $\dfrac{0.6}{0.3}$ (g) $\dfrac{4}{0.05}$ (h) $\dfrac{0.15}{0.03}$ (i) $\dfrac{3}{0.06}$ (j) $\dfrac{28}{0.07}$

C3 These three cards  can be arranged to make a division: $\dfrac{4}{0.5}$ = 8

Arrange these six cards to make two correct divisions.

0.3   = 0.5   1.5   30 $\dfrac{}{15}$ = 5

C4 Arrange these six cards to make two correct divisions.

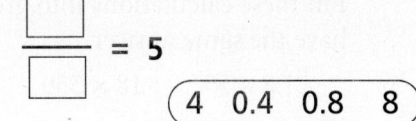

0.8   0.24   2.4   8   = 3   = 0.03

C5 Choose two of the four numbers to go in the boxes to make a correct division.

$$\dfrac{\square}{\square} = 5$$

(4   0.4   0.8   8)

C6 Do the same for each of these.

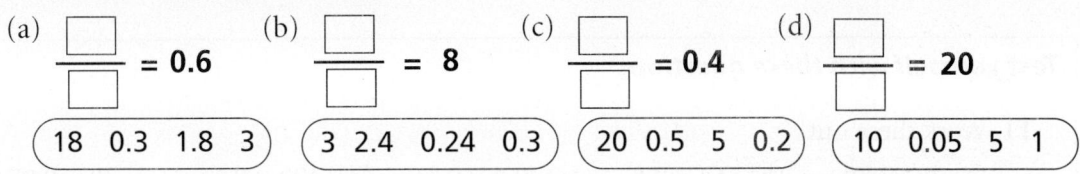

(a) $\dfrac{\square}{\square}$ = 0.6   (b) $\dfrac{\square}{\square}$ = 8   (c) $\dfrac{\square}{\square}$ = 0.4   (d) $\dfrac{\square}{\square}$ = 20

(18 0.3 1.8 3)   (3 2.4 0.24 0.3)   (20 0.5 5 0.2)   (10 0.05 5 1)

Ⓓ *Written division*

Example	$\dfrac{59.8}{0.26}$

Get rid of decimals on the bottom. Here we multiply top and bottom by 100: $$\dfrac{59.8}{0.26} = \dfrac{5980}{26}$$	Get a rough idea of the answer, by rounding: $$\dfrac{6000}{30} = 200$$	Divide 5980 by 26. Here is one way. The result is **230**. $$\begin{array}{r} 230 \\ 26)\overline{5980} \\ 52 \\ \hline 78 \\ 78 \end{array}$$

D1 Work these out.

(a) $\dfrac{32.2}{1.4}$ (b) $\dfrac{7.92}{2.4}$ (c) $\dfrac{2.24}{0.35}$ (d) $\dfrac{6.48}{0.72}$ (e) $\dfrac{11.2}{0.14}$

(f) $\dfrac{44.2}{0.26}$ (g) $\dfrac{1.69}{0.13}$ (h) $\dfrac{5.2}{0.65}$ (i) $\dfrac{0.54}{1.8}$ (j) $\dfrac{34.2}{1.8}$

D2 You are told that $\dfrac{2482}{34}$ = 73. Write down the answer to (a) $\dfrac{2.482}{3.4}$ (b) $\dfrac{248.2}{3.4}$ (c) $\dfrac{24.82}{0.34}$

E Mixed questions

E1 You are told that $17 \times 14 = 238$.

Write down the answers to these.

(a) 1.7×1.4 (b) 170×1.4 (c) 17×0.14 (d) $238 \div 14$ (e) $2380 \div 14$

(f) $238 \div 17$ (g) $23.8 \div 1.7$ (h) $2.38 \div 1.7$ (i) $0.238 \div 1.7$ (j) $23.8 \div 0.14$

E2 Given that $29 \times 43 = 1247$, write down the answers to these.

(a) 290×430 (b) 2.9×0.43 (c) 0.29×43 (d) 0.29×0.43 (e) $1247 \div 2.9$

(f) $124.7 \div 29$ (g) $124.7 \div 2.9$ (h) $12.47 \div 43$ (i) $12\,470 \div 4.3$ (j) $1.247 \div 0.43$

E3 You are told that $18 \times 55 = 990$.

Put these calculations into groups, so that the calculations in each group have the same answer.

1.8×55 18×550 0.18×550 180×55 0.18×0.55

18×0.55 0.18×55 1.8×5.5 180×0.55

Test yourself with these questions

T1 Work these out.

(a) 0.4×80 (b) 0.4×0.8 (c) 0.4×0.2 (d) 400×0.6 (e) 30×0.07

T2 Work these out.

(a) 1.4×23 (b) 1.4×0.23 (c) 2.6×1.8 (d) 260×0.18 (e) 45×0.32

T3 Work these out.

(a) $\dfrac{2.4}{4}$ (b) $\dfrac{240}{400}$ (c) $\dfrac{2.4}{0.4}$ (d) $\dfrac{0.18}{3}$ (e) $\dfrac{1.8}{0.03}$

T4 Work these out.

(a) $\dfrac{2.08}{1.3}$ (b) $\dfrac{64.8}{0.27}$ (c) $\dfrac{5.44}{0.34}$ (d) $\dfrac{0.364}{1.3}$ (e) $\dfrac{11.2}{0.35}$

T5 Given that $63 \times 82 = 5166$, write down the answers to these.

(a) 630×8.2 (b) 6.3×0.82 (c) 0.63×8.2 (d) 0.63×820 (e) 0.63×0.82

(f) $51.66 \div 63$ (g) $516.6 \div 8.2$ (h) $5.166 \div 0.82$ (i) $516.6 \div 820$ (j) $51\,660 \div 820$

5 Linear equations

You should know that, for example, $3(x + 4) = 3x + 12$

$$\frac{3x + 6}{3} = x + 2$$

You will

◆ Simpify expressions such as $3 - 6x + 4x$

◆ Solve a variety of linear equations, including those where you need to simplify first

◆ Form equations to solve problems

A Solving simple equations

Examples

Each expression in brackets shows what is done to **both** sides to get the next line.

$3x - 1 = x + 13$	$[+ 1]$
$3x = x + 14$	$[- x]$
$2x = 14$	$[\div 2]$
$x = 7$	

$\frac{x}{3} + 5 = 9$	$[- 5]$
$\frac{x}{3} = 4$	$[\times 3]$
$x = 12$	

$5x - 8 = 20 - 2x$	$[+ 2x]$
$7x - 8 = 20$	$[+ 8]$
$7x = 28$	$[\div 7]$
$x = 4$	

A1 Solve

(a) $5x = 35$

(b) $\frac{y}{6} = 5$

(c) $4z + 9 = 21$

(d) $\frac{k}{4} + 8 = 13$

(e) $10h - 3 = 47$

(f) $\frac{m}{2} - 1 = 19$

(g) $40 - y = 30$

(h) $10 - 3g = 4$

(i) $20 - 6n = 2$

A2 Solve

(a) $2n = 3$

(b) $4n = 5$

(c) $2n + 3 = 4$

(d) $6n - 4 = 5$

(e) $5n - 3 = 3$

(f) $10n - 1 = 5$

(g) $9 - 4n = 3$

(h) $3 - 2n = 0$

(i) $6 - 10n = 2$

A3 Solve

(a) $x + 9 = 4x$

(b) $x + 5 = 2x + 4$

(c) $4x + 3 = 2x + 13$

(d) $3x + 5 = 7x - 3$

(e) $4x + 6 = 8x - 6$

(f) $5x - 9 = 3x - 1$

(g) $2x + 5 = 6x - 1$

(h) $6x + 5 = 2x + 7$

(i) $2x - 1 = 4x - 6$

A4 Solve

(a) $3n + 6 = 10 - n$

(b) $12 - 2n = 3n + 2$

(c) $15 - 3n = 4n - 6$

(d) $2n - 5 = 15 - 3n$

(e) $8 - n = 15 - 2n$

(f) $15 - 2n = 20 - 3n$

(g) $3n + 1 = 7 - n$

(h) $6n - 13 = 12 - 4n$

(i) $3 - n = 5 - 5n$

B Simplifying

Examples

$9 + 2a - 3 - 7a$	$1 - 9n + 10 + 4n$	$8 - 3n - 5 - 4n - 1$
$= 9 - 3 + 2a - 7a$	$= 1 + 10 - 9n + 4n$	$= 8 - 5 - 1 - 3n - 4n$
$= 6 - 5a$	$= 11 - 5n$	$= 2 - 7n$

B1 Simplify these expressions.

(a) $5 + 3m + 2 - 5m$ (b) $7n - 1 - 4n - 3$ (c) $12 + 4p + 2 - 9p$

(d) $3q + 10 - 7q + 2$ (e) $2 - 8v + 3v - 1$ (f) $8 + w - 3 - 3w - w$

(g) $6 - 3g + 3 - 8g - 2$ (h) $4 - 2h - 3 - h$ (i) $4 - 7k + 5 - 4k + 2k$

B2 Find and simplify expressions for the perimeters of these shapes.

(a)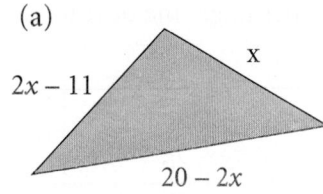

(b)

(c)

C Shape up

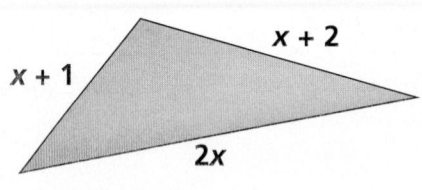

What value of x

- gives a triangle with a perimeter of 63?
- gives a triangle and rectangle with equal perimeters?
- makes the rectangle into a square?

C1 (a) (i) Make a sketch of the triangle for $x = 3$.

      (ii) Work out its perimeter.

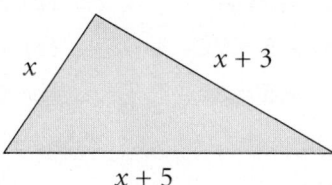

(b) Find an expression for the perimeter of the triangle.

(c) What value of x gives a perimeter of 50?

C2 (a) (i) Make a sketch of the triangle for $x = 1$.

(ii) What is the special name for your triangle?

(iii) Work out its perimeter.

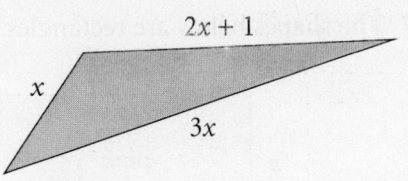

(b) Find an expression for the perimeter of the triangle.

(c) What value of x gives a perimeter of 55?

C3 (a) Find the length and width of the rectangle when $x = 5$.

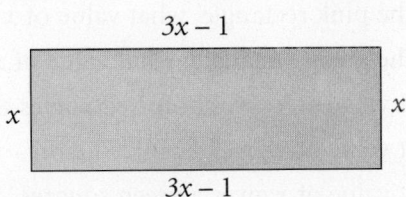

(b) Find an expression for the perimeter of the rectangle.

(c) What value of x gives a perimeter of 10?

(d) What value of x gives a square?

C4 (a) Find an expression for the perimeter of the rectangle.

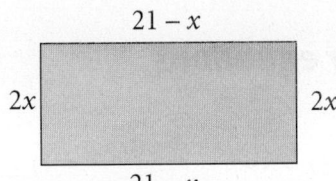

(b) What value of x gives a perimeter of 48?

(c) What value of x gives a square?

C5 (a) Find an expression for the perimeter of the kite.

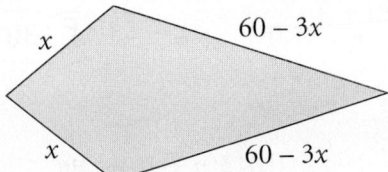

(b) What value of x gives a perimeter of 68?

C6 (a) Find the perimeter of each triangle below when $x = 3$.

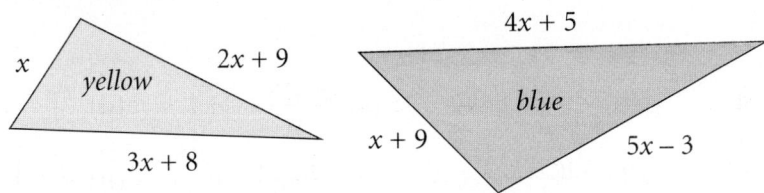

(b) (i) Find an expression for the perimeter of the yellow triangle.

(ii) What value of x gives a yellow triangle with a perimeter of 143?

(c) (i) Find an expression for the perimeter of the blue triangle.

(ii) What value of x gives a blue triangle with a perimeter of 50?

(d) What value of x gives both triangles the same perimeter?

C7 The shapes below are rectangles.

$30 - 2x$

| x | pink |

$42 - 3x$

| x | green |

(a) Find the length and width of each rectangle when $x = 13$.

(b) Explain why it is not possible to sketch a green rectangle for $x = 14$.

(c) For the pink rectangle, what value of x gives a perimeter of 44?

(d) For the green rectangle, what value of x gives a perimeter of 60?

(e) What value of x gives both rectangles the same perimeter?

(f) What value of x gives a pink square?

(g) What value of x gives a green square?

*(h) Explain why it is not possible to sketch a pink rectangle with a perimeter of 100.

Ⓓ *Harder equations*

Ⓐ $2(n + 3) = 10$

Ⓑ $6(2n - 1) = 10n$

Ⓒ $\dfrac{4n + 6}{2} = 3n$

Ⓓ $\dfrac{n - 5}{19} = 7$

Ⓔ $\dfrac{4n + 3}{5} = n$

Ⓕ $3(n + 4) = 6(n - 1)$

D1 Solve

(a) $2(m + 3) = 40$

(b) $3(n + 10) = 9n$

(c) $4(p + 9) = 10p$

(d) $3(2g - 1) = 24$

(e) $5(2h - 3) = 5h + 10$

(f) $4(3k - 7) = 2(3k - 2)$

(g) $2(x + 15) = 3(x + 1)$

(h) $4(y + 9) = 12y - 4$

(i) $5(z - 3) = 7(z - 5)$

(j) $2(a + 5) = 4a + 7$

(k) $3(2b - 2) = 2(b + 2)$

(l) $4(c - 2) = 5(2c - 7)$

D2 Solve

(a) $\dfrac{n - 1}{5} = 4$

(b) $\dfrac{3n + 6}{3} = 10$

(c) $\dfrac{n + 5}{6} = 11$

(d) $\dfrac{4n}{7} - 1 = 7$

(e) $\dfrac{n + 12}{4} = n$

(f) $\dfrac{2n + 132}{13} = n$

(g) $\dfrac{8n - 3}{7} = n$

(h) $\dfrac{n + 18}{2} = 5n$

D3 Solve

(a) $4(10 - n) = 24$

(b) $3(2 - p) = p - 14$

(c) $5(20 - 3q) = 10$

(d) $6(8 - u) = 3(2u - 4)$

(e) $5(10 - 3a) = 4(2a + 1)$

(f) $10(w + 3) = 18(4 - w)$

(g) $3(5 - 2v) = 9(2 - v)$

(h) $4(10 - z) = 3(11 - z)$

(i) $2(11 - 2h) = 3(14 - 3h)$

D4 Solve

(a) $\dfrac{10 - n}{4} = 2$ (b) $\dfrac{24 - n}{9} = 2$ (c) $\dfrac{15 - 3n}{2} = 3$ (d) $\dfrac{39 - 2n}{8} = 4$

D5 Solve

(a) $6(5 - k) = 8(k + 2)$ (b) $\dfrac{8 - 4k}{2} = 1$ (c) $4(k + 2) = 6(5 - 3k)$

(d) $\dfrac{20 - k}{3} = 6$ (e) $8(k - 2) = 17 - 3k$ (f) $\dfrac{18 + k}{3} = k$

(g) $2(11 - 4k) = 7 - 2k$ (h) $3 - 2k = 6 - 4k$ (i) $3(6 - k) = 4(11 - 4k)$

Solving harder equations

Examples

> It is often a good idea to multiply to get rid of brackets as soon as you can.

$$\begin{aligned}
6(x + 2) &= 3(13 - x) \quad \text{[multiply out brackets]}\\
6x + 12 &= 39 - 3x \quad \text{[+ 3x]}\\
9x + 12 &= 39 \quad \text{[−12]}\\
9x &= 27 \quad \text{[÷ 9]}\\
x &= 3
\end{aligned}$$

> It is often a good idea to multiply to get rid of divisions as soon as you can.

$$\begin{aligned}
\dfrac{4n - 5}{2} &= n \quad \text{[× 2]}\\
4n - 5 &= 2n \quad \text{[+ 5]}\\
4n &= 2n + 5 \quad \text{[− 2n]}\\
2n &= 5 \quad \text{[÷ 2]}\\
n &= 2.5
\end{aligned}$$

Test yourself with these questions

T1 Solve these equations.

(a) $4x - 1 = 39$ (b) $3x + 14 = 5x$ (c) $5x + 1 = 7x - 3$

(d) $4x + 1 = 10 - 2x$ (e) $16 - x = 10 + 2x$ (f) $10 - 2x = 16 - 5x$

T2 The shape is a rectangle.

$15 - n$

$2n$

(a) Make a sketch of the rectangle when $n = 2$.

(b) (i) Find an expression for the perimeter of the rectangle.

   (ii) What value of n gives a perimeter of 41?

T3 Solve

(a) $6(k + 5) = 33$ (b) $3(2k - 5) = k$ (c) $2(k + 1) = 10(k - 1)$

(d) $\dfrac{k + 5}{3} = 4$ (e) $\dfrac{2k + 9}{5} = k$ (f) $5(3 - k) = 2(7 - 2k)$

6 Negative numbers

You should know how to substitute in expressions such as $x^2 + 1$, $3(x + 4)$, $\frac{x}{3} - 1$; the rules about the order of calculating with $+$, $-$, $\times$, $\div$ and brackets, and how to solve equations.

You will learn how to calculate with positive and negative numbers and how to substitute in a variety of expressions.

A Calculating

Calculating with positive and negative numbers	Examples
• To add a negative, subtract the corresponding positive	$6 + {}^-9 = 6 - 9$ $= {}^-3$ ${}^-7 + {}^-1 = {}^-7 - 1$ $= {}^-8$
• To subtract a negative, add the corresponding positive	$6 - {}^-9 = 6 + 9$ $= 15$ ${}^-7 - {}^-1 = {}^-7 + 1$ $= {}^-6$
• Multiplying or dividing two negatives gives a positive	${}^-8 \times {}^-2 = 16$ ${}^-14 \div {}^-7 = 2$
• Multiplying or dividing a negative and a positive gives a negative	$5 \times {}^-3 = {}^-15$ ${}^-12 \div 4 = {}^-3$

A1 (a) Copy and complete the addition below using three different numbers from the loop.

$$\boxed{} + \boxed{} = \boxed{}$$

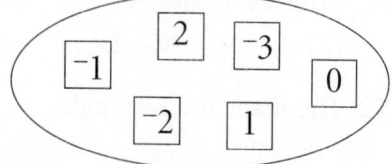

2 ${}^-3$ ${}^-1$ 0 ${}^-2$ 1

(b) Make as many different additions like this as you can.

(c) Copy and complete the subtraction here using three different numbers from the loop.

$$\boxed{} - \boxed{} = \boxed{}$$

(d) Make eight different subtractions like this.

A2 Copy and complete each multiplication table.

(a)
×	${}^-2$
${}^-5$	
4	12

(b)
×	
${}^-1$	${}^-5$
6	${}^-18$

(c)
×	${}^-7$
8	28
6	

A3 This is a 'division triangle'.

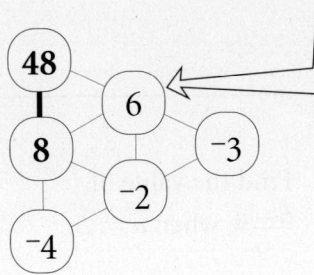

> To find each number, divide the left hand pair of numbers
> top number ÷ bottom number
> (48 ÷ 8 = 6)

Copy and complete each division triangle.

(a)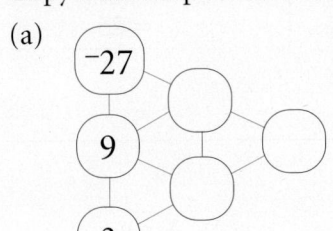

(b)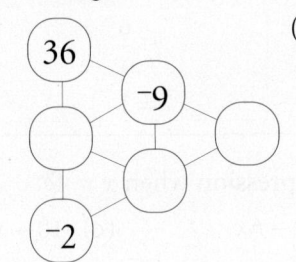

(c)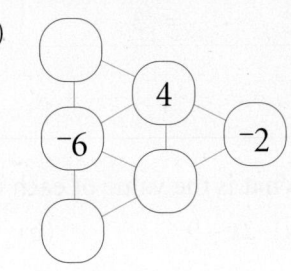

A4 Copy and complete each magic square, so that each row, column and diagonal adds to give the same total.

(a)

		0
	1	
2	⁻3	

(b)

⁻4	1	0
⁻2	⁻3	

(c)

	5	
	⁻3	
3	⁻11	

A5 Find the next two numbers in each sequence.

(a) 11, 9, 7, 5, 3, …

(b) ⁻11, ⁻8, ⁻5, ⁻2, …

(c) 5, 3.5, 2, 0.5, …

(d) 1, ⁻2, 4, ⁻8, 16, …

A6 Work out

(a) $10 + (^-3 \times 2)$

(b) $^-3 \times (^-2 + ^-5)$

(c) $2 \times (1 - 9)$

(d) $^-2 \times 5 - 3$

(e) $6 - ^-4 \times ^-2$

(f) $7 - (1 - 5)$

(g) $\frac{^-12}{3} + 1$

(h) $^-5 + \frac{^-16}{^-4}$

(i) $\frac{^-8 - 2}{5}$

(j) $\frac{^-10 - ^-8}{^-2}$

(k) $\frac{20}{^-5} - ^-3$

(l) $^-2 - \frac{^-15}{5}$

*__A7__ Find the missing number in each of these calculations.

(a) $^-5 \times 3 \times \blacksquare = 30$

(b) $6 + 2 \times \blacksquare = 4$

(c) $(\blacksquare - 5) \times 3 = ^-9$

(d) $4 - ^-2 \times \blacksquare = 14$

(e) $\blacksquare - ^-3 \times ^-2 = ^-4$

(f) $10 + \blacksquare \times ^-5 = 0$

(g) $\frac{10 - \blacksquare}{^-4} = ^-3$

(h) $\frac{\blacksquare}{^-3} - 9 = ^-1$

(i) $^-1 - \frac{10}{\blacksquare} = 4$

B Substitution

Examples

Find the value of
$4x - 7$ when $x = 1$.

$$4x - 7 = 4 \times 1 - 7$$
$$= 4 - 7$$
$$= {}^-3$$

Find the value of
$\dfrac{6 - a}{9}$ when $a = {}^-3$.

$$\frac{6 - a}{9} = \frac{6 - {}^-3}{9}$$
$$= \frac{9}{9}$$
$$= 1$$

Find the value of
$^-5(n - 7)$ when $n = 2$.

$$^-5(n - 7) = {}^-5 \times (2 - 7)$$
$$= {}^-5 \times {}^-5$$
$$= 25$$

B1 What is the value of each expression when $x = 4$?

(a) $2x - 9$ (b) $19 - 6x$ (c) $10 - x^2$ (d) $\dfrac{x}{^-2} + 8$

B2 Find the value of each expression when $y = {}^-3$.

(a) $2y + 1$ (b) $y^2 + 5$ (c) $8 - y$ (d) $\dfrac{y}{^-3}$

B3 What is the value of each expression when $p = {}^-6$?

(a) $3p - 2$ (b) $2(p + 5)$ (c) $1 - 2p$ (d) $p^2 - 9$

(e) $\dfrac{p}{3} - 1$ (f) $(p + 3)^2$ (g) $5 - \dfrac{p^2}{4}$ (h) p^3

Examples

Find the value of
$4(x^2 - 7)$ when $x = 2$.

$$4(x^2 - 7) = 4 \times (2^2 - 7)$$
$$= 4 \times (4 - 7)$$
$$= 4 \times {}^-3$$
$$= {}^-12$$

Find the value of
$\dfrac{6 - 3h}{^-2}$ when $h = {}^-4$.

$$\frac{6 - 3h}{^-2} = \frac{6 - 3 \times {}^-4}{^-2}$$
$$= \frac{6 - {}^-12}{^-2}$$
$$= \frac{18}{^-2}$$
$$= {}^-9$$

Find the value of
$4a^2 - 38$ when $a = {}^-3$.

$$4a^2 - 38 = 4 \times ({}^-3)^2 - 38$$
$$= 4 \times 9 - 38$$
$$= 36 - 38$$
$$= {}^-2$$

B4 What is the value of each expression when $h = 5$?

(a) $5(2h - 11)$ (b) $\dfrac{10 - 6h}{5}$ (c) $3(10 - h^2)$ (d) $\dfrac{4h}{^-2} + 3$

B5 What is the value of each expression when $k = {}^-4$?

   (a) $2k^2 + 1$ (b) $3(2k + 5)$ (c) $5(2 - 3k)$ (d) $\frac{1}{2}k^2 - 1$

   (e) $\frac{8 - k^2}{-4}$ (f) $\frac{k^2}{8} - 10$ (g) $\frac{(k - 2)^2}{9}$ (h) $5 - \frac{k^2}{2}$

B6

$\dfrac{3n - 17}{2}$	$5 - n^2$	$3n^2 - 11$	$\dfrac{4n + 3}{-5}$	$2(n - 5)$	$2(n^2 - 7)$

   (a) When $n = 3$, three expressions above have a value of $^-4$. Find these expressions.

   (b) When $n = {}^-2$, three expressions above have the same value. Find these expressions.

B7 Sixteen expressions are arranged in a square grid giving four horizontal, four vertical and two diagonal sets of expressions.

For example, a set of expressions is shaded on the grid.

$\dfrac{n - 3}{2}$	$2n$	$n^2 - 3$	$n - 1$
$\dfrac{2n^2}{-6}$	$4(12 - n)$	$\dfrac{6 - n}{4}$	$3n + 11$
$4n + 9$	$\dfrac{n^2}{4}$	$\dfrac{5 - n}{7}$	$2n + 5$
$\dfrac{4n - 3}{5}$	$n + 8$	$3n^2$	$\dfrac{n^2}{2} - 25$

   (a) What is the value of each expression on a shaded square when $n = 8$? What do you notice?

   (b) When $n = 2$, find a set of expressions (in a row, column or diagonal) that each have a value of 1.

   (c) When $n = {}^-1$, which set of expressions each have a value of $^-2$?

   (d) Find sets where each expression has the same value when

      (i) $n = {}^-2$ (ii) $n = {}^-3$ (iii) $n = {}^-6$

B8 (a) Find the value of the expression $4n - 3$ when

      (i) $n = 1$ (ii) $n = 0$ (iii) $n = {}^-2$

   (b) Find the value of the expression $3n - 18$ when

      (i) $n = 1$ (ii) $n = 0$ (iii) $n = {}^-2$

   (c) There is one value of n for which both expressions have the same value.

      (i) By putting $4n - 3$ equal to $3n - 18$ and solving the equation, find this value of n.

      (ii) Check that the value is correct by substituting it in both expressions.

C Equations with negative numbers

C1 (a) Solve the equation $2n - 4 = {}^-18$.

(b) Check that your answer fits the original equation.

C2 Solve the equation $3m - 5 = 4m + 16$. Check your answer works.

C3 Solve each of these equations. Check each of your answers.

(a) $3k + 6 = 4k + 10$ (b) $7g + 2 = 3g - 10$ (c) $8 - n = 12 + n$

(d) $12 - 3b = 8 - 4b$ (e) $4t + 28 = 10 - 2t$ (f) $\dfrac{x}{2} - 1 = x + 1$

C4 Solve and check each of these.

(a) $2(r + 8) = 4$ (b) $3(x - 2) = 4(x + 3)$ (c) $6(3 - q) = 4q + 33$

(d) $\dfrac{2a + 5}{2} = 2$ (e) $\dfrac{9 - 3m}{5} = 6$ (f) $\dfrac{20 + 9k}{5} = k$

C5 Write equations for each of these number puzzles and solve them.

(a)

I think of a number.

I double it.

I add 11.

My answer is 3.

(b)

I think of a number.

I multiply it by 3.

I add 20.

My answer is the same as the number I started with.

(c)

I think of a number.

I double it.

I take my answer away from 10.

My answer is 19 more than the number I started with.

Test yourself with these questions

T1 Evaluate

(a) $^-3 - 6$ (b) $^-3 \times {}^-4$ (c) $4 + {}^-8$

(d) $^-3 \times (2 - 9)$ (e) $\dfrac{^-7 + {}^-8}{^-5}$ (f) $7 - \dfrac{^-8}{4}$

T2 What is the value of each expression when $n = {}^-8$?

(a) $2n - 10$ (b) $5(n + 1)$ (c) $\dfrac{3n}{4} + 12$ (d) $\dfrac{n^2 + 1}{5}$

(e) $^-3(n - 7)$ (f) $2n^2 - 100$ (g) $\dfrac{20 - 2n}{^-9}$ (h) $3 - \dfrac{n}{4}$

T3 Solve each of these equations and check that your answers work.

(a) $4x + 11 = 2x + 3$ (b) $5(k + 2) = {}^-40$ (c) $2m - 13 = 5m - 22$

(d) $3w + 5 = 7(w + 3)$ (e) $\dfrac{3u + 5}{7} = {}^-10$ (f) $3(d + 6) = d + 2$

Review 1

Do not use a calculator for these questions.

1 This diagram shows a design for a square patchwork cushion.

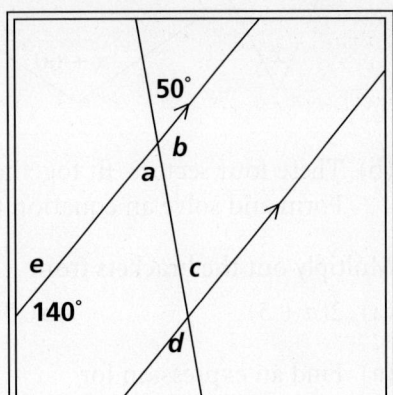

(a) Copy and complete
'Angles *a* and ___ are alternate angles.'

(b) Find a pair of angles that are vertically opposite.

(c) Work out the size of each lettered angle. Give a reason each time.

2

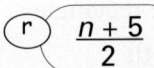

$\boxed{p}\; 3(n - 2)$ $\boxed{q}\; 2(3n - 5)$ $\boxed{r}\; \dfrac{n + 5}{2}$ $\boxed{s}\; 7 - 2n$

(a) Find the value of each expression when   (i) $n = 4$ (ii) $n = {}^-1$

(b) Which expression has the smallest value when $n = 0$?

3 Work out these.

(a) ${}^-3 \times (1 - 7)$ (b) $5 - \dfrac{{}^-16}{8}$ (c) $\dfrac{{}^-3 - 9}{{}^-4}$

4 Which expression below gives the size of the angle marked with a dot?

$\boxed{x + 180}$ $\boxed{x - 180}$

$\boxed{180 - x}$ $\boxed{x + 90}$

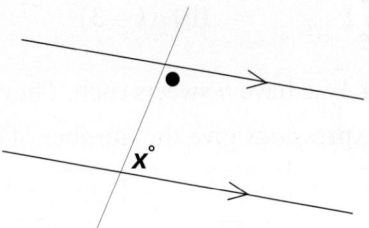

5 Simplify these expressions.

(a) $4x \times 5$ (b) $5 + 3p - 1 - 5p$ (c) $\frac{1}{4}(8y - 20)$

(d) $\dfrac{10b}{5} + 1$ (e) $\dfrac{6w + 15}{3}$ (f) $\dfrac{12 - 18k}{6}$

6 Find the area of each rectangle in m².

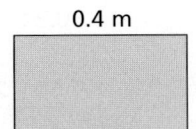

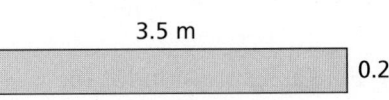

7 How many pieces of ribbon 0.7 m long could be cut from a roll of ribbon 28 m long?

8 (a) Write and simplify an expression for the sum of
the four angles marked in the sectors below.

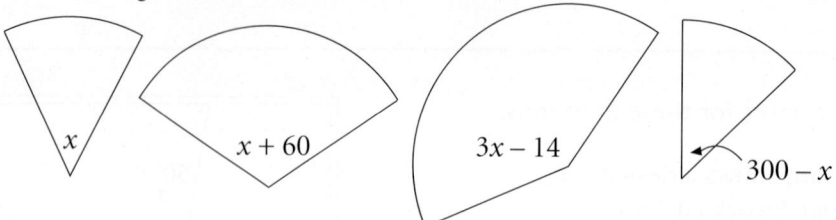

x $x + 60$ $3x - 14$ $300 - x$

(b) These four sectors fit together to make a complete circle.
Form and solve an equation to find the value of x.

9 Multiply out the brackets from

(a) $2(n + 5)$ (b) $3(5 - 2m)$ (c) $7(3p - 5)$

10 (a) Find an expression for
the perimeter of
(i) the triangle   (ii) the rectangle

(b) What value of x gives both
shapes the same perimeter?

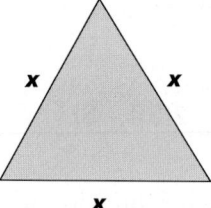

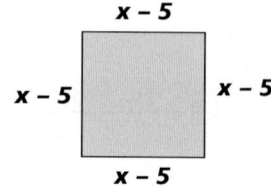

11 Calculate 0.03×0.15

12 Solve these equations.

(a) $3x + 5 = 5x - 6$ (b) $7x + 9 = 3x + 1$ (c) $5x - 18 = 3 - 2x$

13 What is the value of each expression when $k = {}^-5$?

(a) $k^2 + 1$ (b) $(k - 3)^2$ (c) $\dfrac{1 - 3k}{-2}$ (d) $\dfrac{k - 3}{5}$

14 Jane and Ayse have n sweets each. They are each given 2 sweets.

Which expressions give the number of sweets they have altogether?

$n + 2$		$2n + 2$		$2n + 4$		$2(n + 4)$		$2(n + 2)$

15 Solve these equations.

(a) $5(n - 1) = 2n + 4$ (b) $5n + 11 = 3(3n - 5)$

(c) $6(2n + 5) = 9(1 - n)$ (d) $\dfrac{10 - 3n}{2} = n$

16 Calculate   (a) $\dfrac{0.21}{0.07}$ (b) $\dfrac{2.85}{0.19}$ (c) $\dfrac{8.4}{0.24}$ (d) $\dfrac{0.92}{2.3}$

17 (a) Find an expression for the
perimeter of this rectangle.

(b) What is the perimeter when $n = 10$?

(c) What value of n gives a perimeter of 59?

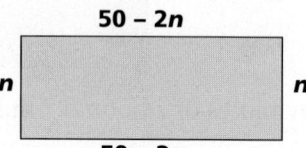

Multiples, factors and powers

You will

◆ work out factors, multiples, prime numbers and powers

◆ use index notation and the rules for multiplying powers of the same number

◆ find and use prime factorisations, for example to work out lowest common multiples and highest common factors

A Multiples, factors and primes

CELL 23

Lock up

A prison has 25 prisoners (one in each cell) and 25 jailers.
The cells are numbered 1 to 25.

The jailers all go to a party one night and return very merry!

• The first jailer unlocks every cell.

• The second jailer locks every cell whose number is a multiple of 2.

• The third jailer turns his key in the lock for cells whose numbers are multiples of 3. (He locks or unlocks these cells).

• The fourth jailer turns his key in the lock for cells whose numbers are multiples of 4

… and so on till the twenty-fifth jailer.

All the jailers then fall asleep and the prison is silent.

The prisoners all try their doors - which ones escape?

Multiples

Multiples of 7 are numbers that can be divided exactly by 7.
Examples are 42, 84, 14, 21 and 7 itself.
Multiples of 3 are numbers that can be divided exactly by 3. Examples are 6, 21, 30, 63.

Common multiples of 3 and 7 are numbers that can be divided exactly by 3 **and** by 7.
Some common multiples of 3 and 7 are 42, 21, 63 and 105

21 is the **lowest common multiple** of 3 and 7.

A1 Which numbers in this list are multiples of 6?
36, 6, 3, 16, 12, 60, 600, 80

A2 (a) Write down five multiples of 4.

(b) Write down five multiples of 3.

(c) Write down a common multiple of 4 and 3.

Factors

Factors of 20 are numbers that divide exactly into 20.
They can be found in pairs.

For example: $5 \times 4 = 20$ so $20 \div 5 = 4$ and $20 \div 4 = 5$
So 4 and 5 are factors of 20.

Factor pairs for 20

1 2 4 5 10 20

Factors of 20 are 1, 2, 4, 5, 10 and 20.
Factors of 8 are 1, 2, 4 and 8.

Common factors of 8 and 20 are numbers that divide exactly into 8 **and** 20.
They are 1, 2 and 4.

4 is the **highest common factor** of 8 and 20.

A3 One factor of 32 is missing from this list. Which is it?

16, 4, 2, 1, 32

A4 Which of these numbers are factors of 36?

5, 8, 2, 3, 4, 10, 9, 12

A5 Write down all the factors of 60.

A6 Decide whether each statement is true or false.

(a) 5 is a factor of 10 (b) 6 is a multiple of 12 (c) 14 is a multiple of 7

(d) 8 is a factor of 4 (e) 28 is a multiple of 4 (f) 4 is a factor of 28

A7 (a) Write down three different factors of 10.

(b) List three different multiples of 10.

A8 Which numbers in the loop are:

(a) multiples of 5

(b) factors of 10

(c) multiples of 3

(d) common multiples of 3 and 5

(e) common factors of 12 and 20

(f) common factors of 5 and 12

3 10 4 30 60 2 17 5 15 1

A9 (a) Write down three common multiples of 2 and 8.

(b) Write down a common multiple of 6 and 9.

A10 (a) Find all the common factors of 18 and 45.

(b) What is the highest common factor of 18 and 45?

A11 What is the lowest common multiple of

   (a) 5 and 2   (b) 4 and 8   (c) 12 and 8   (d) 4 and 9

A12 What is the highest common factor of

   (a) 35 and 20   (b) 5 and 9   (c) 10 and 20   (d) 24 and 54

Primes

Prime numbers are numbers with **exactly two** factors.

For example,
17 is a prime number as it has two factors (1 and 17).

Factor pairs for 17

1   17

8 is not a prime number as it has more than two factors (1, 2, 4 and 8).
1 is not a prime number as it only has one factor (1).

A13 Which of the following numbers are not prime?

   2, 6, 14, 19, 11, 9, 13, 21, 10, 15, 5

A14 List all the prime numbers between 20 and 40.

A15 Each set of clues gives a number.
Find these five numbers.

(a)
- A factor of 16
- An odd number

(b)
- Less than 30
- A multiple of 7
- An even number
- A number with six factors

(c)
- A common factor of 10 and 35
- A prime number

(d)
- A common multiple of 4 and 6
- A multiple of 10
- Less than 100

(e)
- A common factor of 40 and 60
- An even number
- A prime number

A16 Both 3 and 11 are prime numbers.

Find the highest common factor of 3 and 11.
Find the lowest common multiple of 3 and 11.

Repeat for some other pairs of prime numbers.
Comment on your results.

B Powers

Ice cream

Toni sells two flavours of ice cream, strawberry and vanilla.

Here are some different ice creams that each have three scoops.

- How many different ice creams can Toni make with three scoops?
- Investigate for two scoops, one scoop, four scoops, …

Rosa sells three flavours of ice cream, strawberry, vanilla and chocolate.

- How many different ice creams can Rosa make with one scoop, two scoops, … ?

Try B1 to B7 without using a calculator.

B1 Write these in shorthand form using indices.

(a) $2 \times 2 \times 2 \times 2 \times 2 \times 2 \times 2 \times 2$ (b) $4 \times 4 \times 4 \times 4 \times 4 \times 4 \times 4 \times 4 \times 4 \times 4 \times 4$

B2 Find the value of

(a) 2^4 (b) 4^3 (c) 3^5 (d) 7^2 (e) 5^3

B3 (a) List all the powers of two between 10 and 50.

(b) What is the value of 'three to the power two'?

B4 There is one cell in a flask in a laboratory.
The number of cells doubles every 15 minutes.

(a) How many cells are there after 2 hours?

(b) How long does it take the number of cells to increase to 2^{10}?

B5 Decide if the following statements are true or false.

(a) $5^2 = 5 \times 2$ (b) $3^2 > 2^3$ (c) $2^6 < 5^2$ (d) $3^4 < 6^2$

B6 Choose the correct symbol, $<$, $>$ or $=$, for each box below.

(a) $4^3 \square 3^4$ (b) $7^2 \square 2^7$ (c) $2^5 \square 5^2$ (d) $9^1 \square 1^9$

B7 Find the missing numbers in these statements

(a) $2^{\square} = 64$ (b) $8^{\square} = 8$ (c) $9^{\square} = 81$ (d) $\square^3 = 1$

B8 Most calculators have a special key for working out powers.

Find this key on your calculator.

(It might look like one of these: or or )

Use this key on your calculator to work out

(a) 2^3 (b) 10^4 (c) 3^6

Check your answers are correct without using this key.

B9 Calculate the value of

(a) 13^4 (b) 2^{12} (c) 1^{23} (d) 5^9 (e) 9^5

B10 Arrange the following numbers in order of size, smallest first:

2^{31}, 7^{10}, 3^{20}, 16^6, 100^2

B11 Which do you think will be larger, 2^{25} or 25^4?
Check with your calculator.

B12 Copy and complete this cross number puzzle.

Across	Down
1 3^8	2 A power of 2
5 21^2	3 A common multiple of 11 and 13
6 A factor of 24	
8 A common factor of 45 and 105	4 5^5
	6 A power of 4
9 A power of 3	7 A multiple of 11
10 A power of 7	8 A multiple of 100

In an expression like 2^5, the raised number '5' is called the index.

Indices (more than one index) are used as a mathematical shorthand.

Examples

$2 \times 2 \times 2 \times 2 \times 2 = 2^5$ $3 \times 3 \times 3 \times 3 = 3^4$

The value of 5^3 is $5 \times 5 \times 5 = 125$

We say 3^4 as 'three to the power four'.

Powers of three can be written as 3^1, 3^2, 3^3, 3^4, ...

                         or   3,   9,   27,   81, ...

C *Multiplying*

C1 Find the missing numbers in these calculations.

(a) $2^3 \times 2^2 = (2 \times 2 \times 2) \times (2 \times 2) = 2^{\square}$

(b) $5^3 \times 5^6 = (5 \times 5 \times 5) \times (5 \times 5 \times 5 \times 5 \times 5 \times 5) = 5^{\square}$

(c) $7^5 \times 7^4 = (7 \times 7 \times 7 \times 7 \times 7) \times (7 \times 7 \times 7 \times 7) = 7^{\square}$

(d) $3^5 \times 3 = 3 \times 3 \times 3 \times 3 \times 3 \times 3 = 3^{\square}$

C2 Write down the numbers missing from these calculations.

(a) $3^2 \times 3^3 = 3^{\square}$ 　　(b) $4^2 \times 4^4 = 4^{\square}$ 　　(c) $8 \times 8^7 = 8^{\square}$

(d) $6^3 \times 6^9 = 6^{\square}$ 　　(e) $2^{\square} \times 2^5 = 2^{11}$ 　　(f) $7^5 \times 7^{\square} = 7^6$

C3 (a) Write down a rule for multiplying powers of the same number. Explain why your rule works.

(b) Using your rule, copy and complete $2^{12} \times 2^5 = 2^{\square}$

C4 Find three pairs of equivalent expressions.

A $\boxed{2^5 \times 2^2}$ 　**B** $\boxed{2^9}$ 　**C** $\boxed{2^9 \times 2}$ 　**D** $\boxed{2^7}$ 　**E** $\boxed{2^{10}}$ 　**F** $\boxed{2^5 \times 2^4}$

C5 Write the answers to these using indices.

(a) $3^4 \times 3^3$ 　　(b) $10^5 \times 10^6$ 　　(c) $4^8 \times 4^4$ 　　(d) $8^4 \times 8$

(e) $2^4 \times 2^2 \times 2^3$ 　　(f) $7 \times 7^9 \times 7^2$ 　　(g) $10 \times 10^9 \times 10$ 　　(h) $9^{20} \times 9^{10}$

C6 Copy and complete

(a) $2^{\square} \times 2^4 = 2^{12}$ 　　(b) $5^2 \times 5^{\square} = 5^8$ 　　(c) $3^{\square} \times 3 = 3^{10}$ 　　(d) $4^3 \times 4^{\square} \times 4^5 = 4^{10}$

C7 Which two of these statements are false?

A $\boxed{2^5 \times 3^4 \times 2^2 = 2^7 \times 3^4}$ 　　　**B** $\boxed{2^2 \times 3^5 = 6^7}$

C $\boxed{5^2 \times 6^3 \times 5^4 \times 6 = 5^6 \times 6^4}$ 　　**D** $\boxed{5^4 \times 3^2 \times 5^5 = 15^{11}}$

C8 Copy and complete

(a) $3^2 \times 5^3 \times 5^4 \times 3^6 = 3^{\square} \times 5^{\square}$ 　　(b) $2 \times 9^2 \times 2^5 \times 9^3 = 2^{\square} \times 9^{\square}$

(c) $4^7 \times 3^{\square} \times 4 \times 3^2 = 3^{10} \times 4^{\square}$ 　　(d) $3^4 \times 11^{\square} \times 3^{\square} \times 11^5 = 3^5 \times 11^8$

C9 Simplify

(a) $10^2 \times 3^4 \times 10^3 \times 3^5$ 　　(b) $2^2 \times 5^3 \times 2^9$ 　　　(c) $5^9 \times 7 \times 7^6 \times 5$

To **multiply** powers of the same number, **add** the indices $(a^m \times a^n = a^{m+n})$.

Example 　　　　　$3^4 \times 3^2$

$= (3 \times 3 \times 3 \times 3) \times (3 \times 3)$

$= 3 \times 3 \times 3 \times 3 \times 3 \times 3$ 　　　$= 3^{4+2}$

$= 3^6$ 　　　　　　　　　　　　$= 3^6$

D Prime factorisation

There are many ways to write 84 as a product of factors.

For example: $84 = 4 \times 21$
$84 = 12 \times 7$
$84 = 2 \times 6 \times 7$
$84 = 2 \times 2 \times 3 \times 7$

$2 \times 2 \times 3 \times 7$ is called the **product of prime factors** or **prime factorisation** of 84.
We can use index notation to write it as $2^2 \times 3 \times 7$.

We can work out that 96 is $2 \times 2 \times 2 \times 2 \times 2 \times 3$ (or $2^5 \times 3$) in different ways.

Factor trees

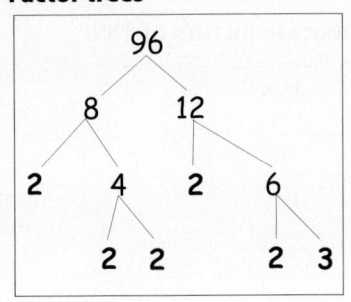

Repeated division

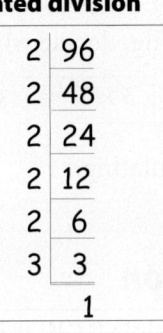

Factor pairs

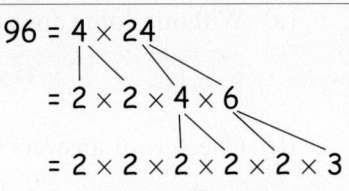

D1 Match each number to its prime factorisation.

A 120   **B** 990   **C** 240   **D** 7425

E $3^3 \times 5^2 \times 11$ **F** $2 \times 3^2 \times 5 \times 11$ **G** $2^4 \times 3 \times 5$ **H** $2^3 \times 3 \times 5$

D2 Find the prime factorisation of each of these numbers and write it using index notation.

(a) 45   (b) 150   (c) 48   (d) 126   (e) 243

D3 The prime factorisation of 875 is $5 \times 5 \times 5 \times 7$.

(a) Without doing any calculating, decide which of these are factors of 875.
Explain how you decided.

| 2 | 3 | 5 | 7 | 20 | 25 | 35 | 45 | 125 |

(b) Check your answers by calculating.

D4 The prime factorisation of 315 is $3 \times 3 \times 5 \times 7$.
The prime factorisation of 3465 is $3 \times 3 \times 5 \times 7 \times 11$.

Without doing any calculating, decide if 3465 is a multiple of 315.
Explain how you decided.

D5 The prime factorisation of 1175 is $3 \times 5 \times 7 \times 11$.
The prime factorisation of 5005 is $5 \times 7 \times 11 \times 13$.

Without doing any calculating, decide if 5005 is a multiple of 1175.
Explain how you decided.

D6 The prime factorisation of 24 is $2^3 \times 3$.

(a) Without doing any calculating, decide which of these are multiples of 24.

| $2^3 \times 3 \times 5$ | $2^3 \times 3 \times 7^2$ | $2^3 \times 5$ | $2^4 \times 3$ | $2^3 \times 3^2$ | $2^2 \times 3 \times 5$ |

(b) Check your answers by calculating.

D7 The prime factorisation of 189 is $3^3 \times 7$.

(a) Without doing any calculating, decide which of these are factors of 189.

| 5 | 7 | $3^3 \times 7^2$ | 3×7 | 3×7^3 | $3^2 \times 7$ |

(b) Check your answers by calculating.

E *Using prime factorisation*

Finding the lowest common multiple (LCM) using prime factorisation

Example: Find the LCM of 15 and 20.

First find the prime factorisation of each number.

$15 = 3 \times 5$
$20 = 2 \times 2 \times 5$

The lowest number which is a multiple of 15 **and** a multiple of 20 is
$2 \times 2 \times 3 \times 5$ which is **60**.

Finding the highest common factor (HCF) using prime factorisation

Example: Find the HCF of 84 and 120.

First find the prime factorisation of each number.

$84 = 2 \times 2 \times 3 \times 7$
$120 = 2 \times 2 \times 2 \times 3 \times 5$

The highest number which is a factor of 84 **and** a factor of 120 is
$2 \times 2 \times 3$ which is **12**.

E1 (a) (i) Find the prime factorisation of 18.

(ii) Find the prime factorisation of 42.

(b) Use your prime factorisations to find the LCM of 18 and 42.

E2 Use prime factorisation to find the LCM of

   (a) 12 and 20 (b) 14 and 15 (c) 45 and 165 (d) 42 and 350

E3 (a) (i) Find the prime factorisation of 64.

   (ii) Find the prime factorisation of 168.

 (b) Use your prime factorisations to find the HCF of 64 and 168.

E4 Use prime factorisation to find the HCF of

   (a) 72 and 180 (b) 90 and 525 (c) 165 and 154 (d) 104 and 234

*E5 Helen wants to make a patchwork quilt from squares. The quilt is to measure 204 cm by 374 cm and the width of each square is to be a whole number of centimetres.

What is the largest size she can use for the squares?

*E6 Ten friends swim at the local pool on 1 January 2000. They make a New Year's resolution.

The first person is going to swim every day, the second person every second day, the third every third day and so on.

How many days later do ten people again swim on the same day?

Test yourself with these questions

T1 Find all the factors of 90.

T2 (a) Write down the prime numbers between 10 and 30.

 (b) Write down all the powers of 3 between 20 and 100.

T3 Evaluate (a) 3^7 (b) 13^2 (c) 10^1

T4 Write the answers to these using indices

 (a) $2^5 \times 2^9$ (b) 3×3^8 (c) $5^2 \times 5^3 \times 5^2$

T5 Find the prime factorisation of 234 and write it using index notation.

T6 (a) Use prime factorisation to find the LCM of 12 and 15.

 (b) Use prime factorisation to find the HCF of 84 and 126.

8 Area and perimeter

This unit will give you practice in calculating

- ◆ the areas of a rectangle, right-angled triangles and shapes made from them
- ◆ the circumference and area of a circle

You will also learn how to

- ◆ find the area of parallelograms, any triangle and trapeziums
- ◆ calculate the radius of a circle given the circumference or area

A Area of a parallelogram

The area of a parallelogram is found by using the formula

Area of a parallelogram = base length × perpendicular height

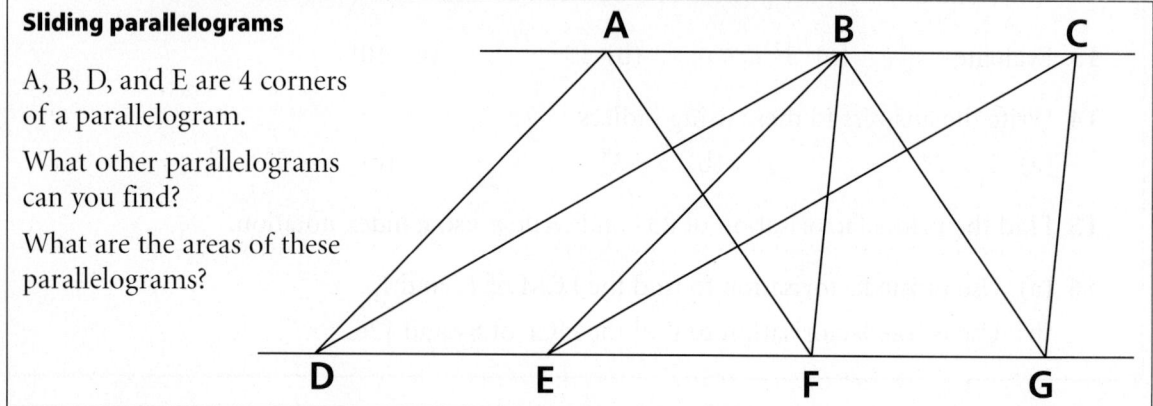

Sliding parallelograms

A, B, D, and E are 4 corners of a parallelogram.

What other parallelograms can you find?

What are the areas of these parallelograms?

A1 Find the area of the parallelograms in this diagram.

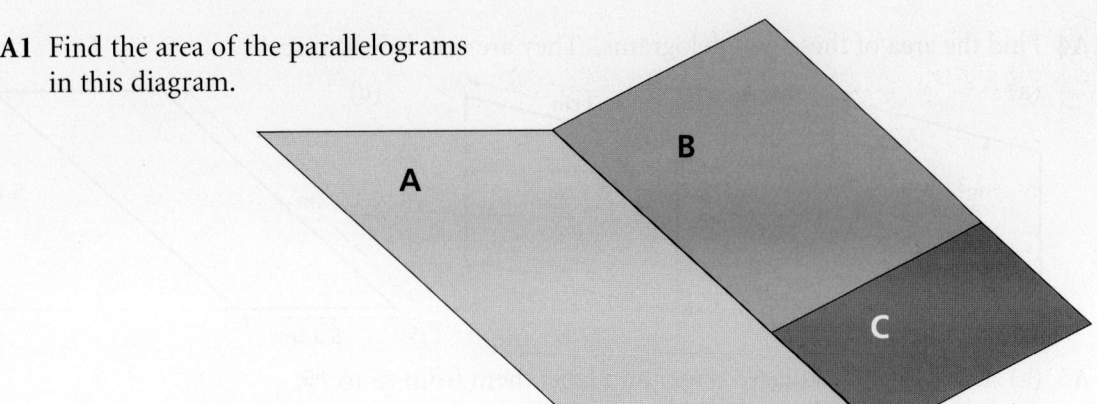

A2 This pattern is made up of 9 different parallelograms.

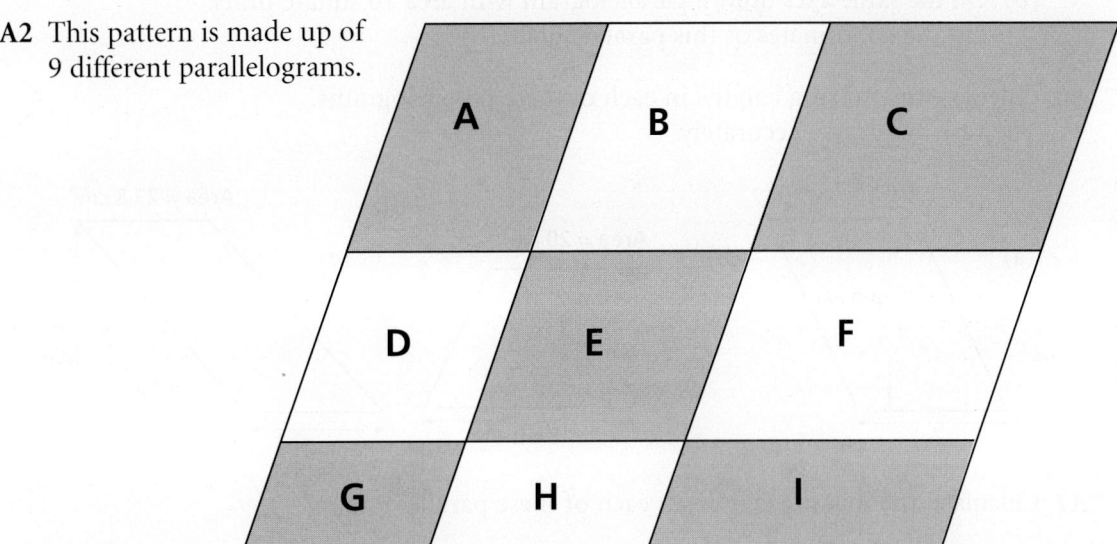

(a) Which two parallelograms have the same area?
What is that area?

(b) List the four pairs of parallelograms where one has twice the area of the other.

A3 (a) Find the area of:

(i) the rectangle around this pattern

(ii) parallelogram A

(iii) parallelogram B

(iv) the area left white

(b) Write down the ratio of the white area to the shaded area in this pattern

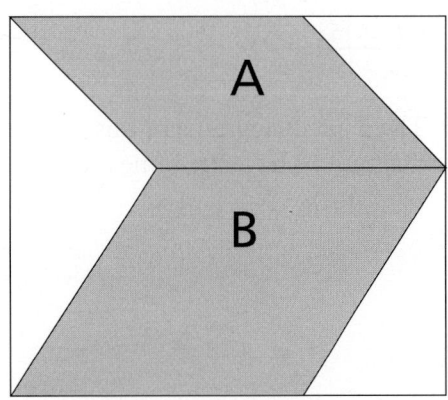

A4 Find the area of these parallelograms. They are not drawn to scale.

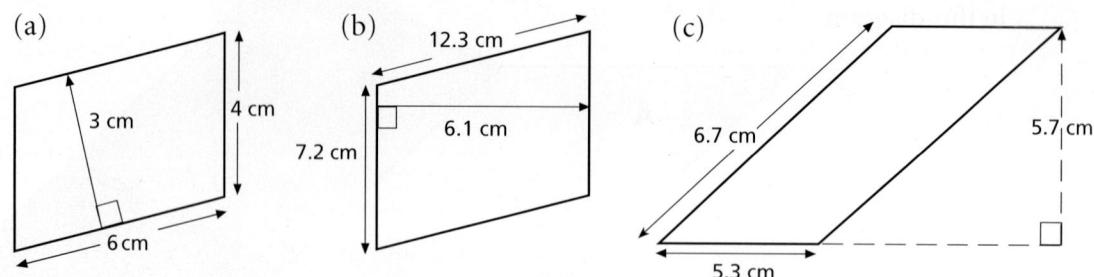

(a) 3 cm, 4 cm, 6 cm

(b) 12.3 cm, 6.1 cm, 7.2 cm

(c) 6.7 cm, 5.7 cm, 5.3 cm

A5 (a) Draw axes on squared paper and label them from ⁻5 to ⁺5.
Draw a parallelogram with vertices at (3, 2), (3, ⁻1), (⁻2, 2), (⁻2, 5).
Calculate its area in square units.

(b) On the same axes draw a parallelogram with area 16 square units.
List the coordinates of this parallelogram.

A6 Calculate the missing lengths in each of these parallelograms.
They are not drawn accurately.

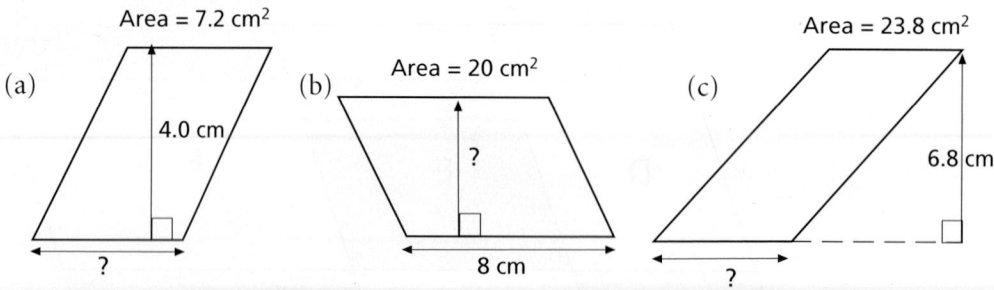

(a) Area = 7.2 cm², 4.0 cm, ?

(b) Area = 20 cm², ?, 8 cm

(c) Area = 23.8 cm², 6.8 cm, ?

*A7 Calculate the missing lengths in each of these parallelograms.

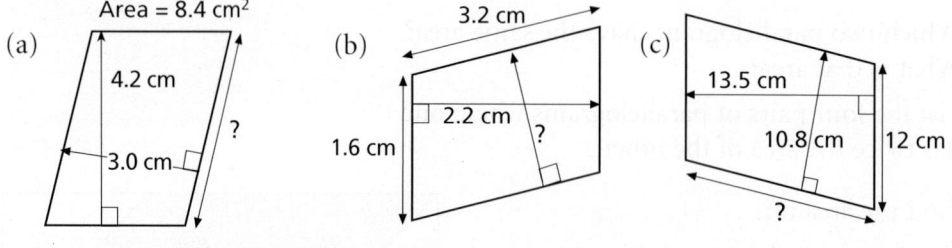

(a) Area = 8.4 cm², 4.2 cm, 3.0 cm, ?

(b) 3.2 cm, 2.2 cm, 1.6 cm, ?

(c) 13.5 cm, 10.8 cm, 12 cm, ?

*A8 The 3 parallelograms in this diagram all have the same area.

Find the missing lengths

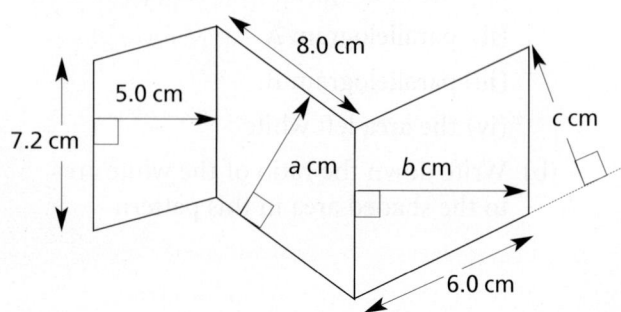

7.2 cm, 5.0 cm, 8.0 cm, a cm, b cm, c cm, 6.0 cm

B *Into triangles*

The area of a triangle is half the area of a parallelogram with the same base and perpendicular height.

To find the area of a triangle use the formula:

Area of a triangle = $\dfrac{\text{base} \times \text{height}}{2}$

or $\quad A = \frac{1}{2}bh \quad$ or $\quad A = \dfrac{bh}{2}$

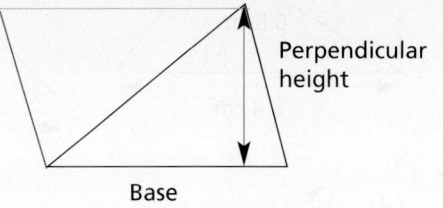

Perpendicular height

Base

B1 Find the area of these shaded triangles by measuring the lengths you need.

(a)

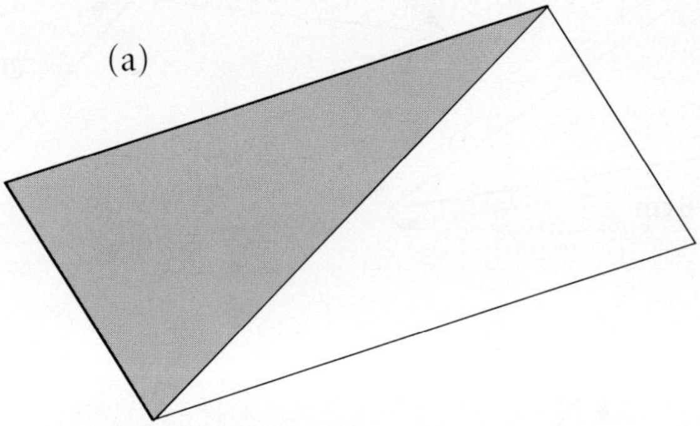

(b)

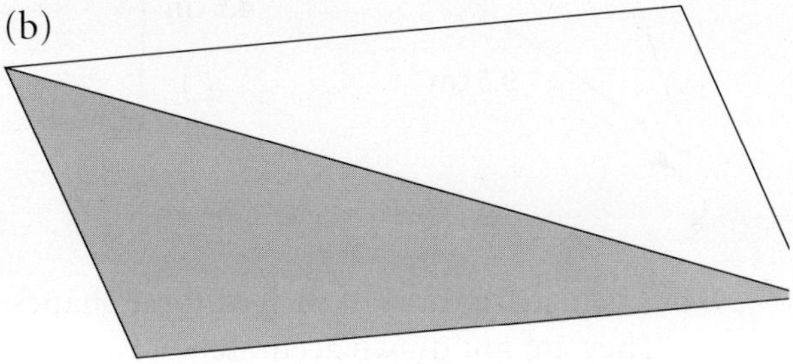

(c)

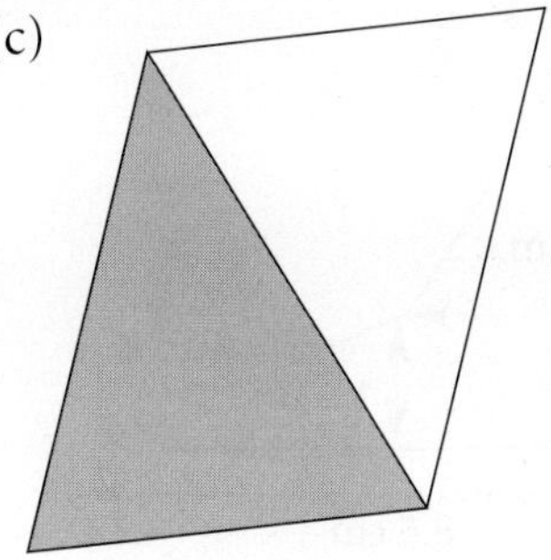

(d)

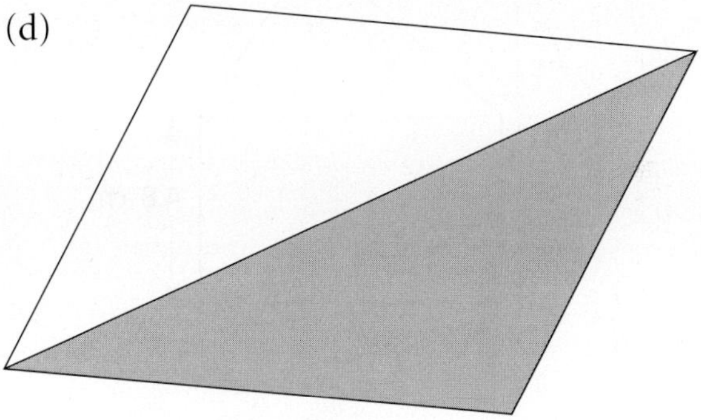

B2 (a) By measuring, find the areas of the triangles A, B and C in this diagram.

(b) Find the area of the surrounding triangle. Check that this is the same as the total area of A, B and C.

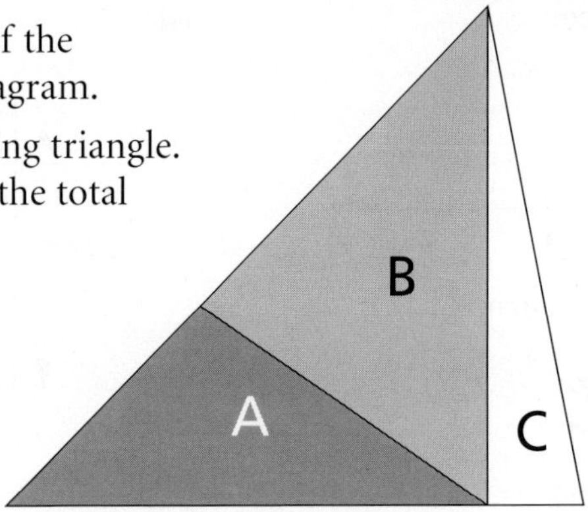

B

A

C

B3 Find the areas of these triangles.
They are not drawn to scale

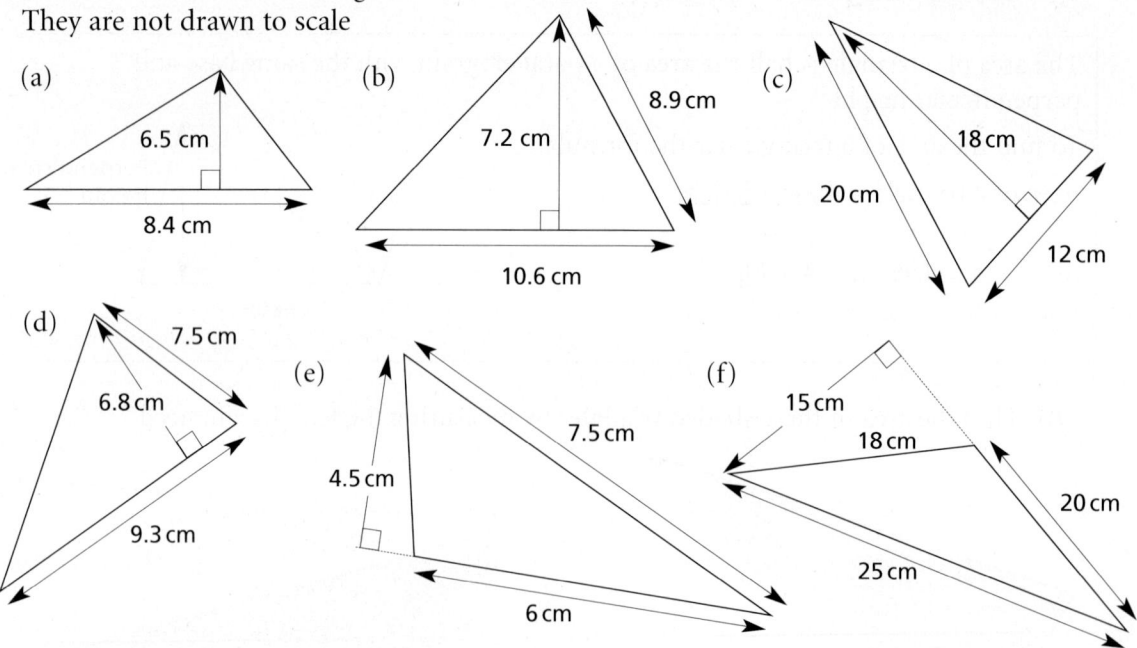

(a) 6.5 cm, 8.4 cm

(b) 7.2 cm, 8.9 cm, 10.6 cm

(c) 18 cm, 20 cm, 12 cm

(d) 7.5 cm, 6.8 cm, 9.3 cm

(e) 4.5 cm, 7.5 cm, 6 cm

(f) 15 cm, 18 cm, 20 cm, 25 cm

B4 Calculate the area of each of these shapes.
They are not drawn accurately.

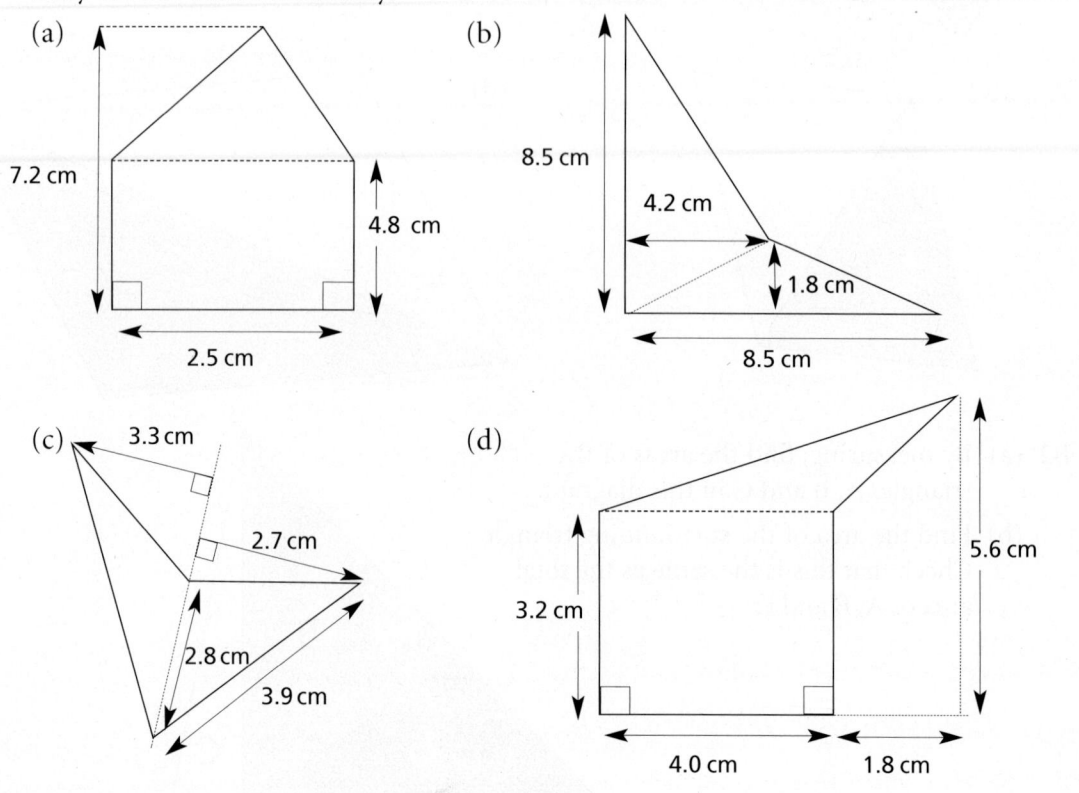

(a) 7.2 cm, 4.8 cm, 2.5 cm

(b) 8.5 cm, 4.2 cm, 1.8 cm, 8.5 cm

(c) 3.3 cm, 2.7 cm, 2.8 cm, 3.9 cm

(d) 5.6 cm, 3.2 cm, 4.0 cm, 1.8 cm

B5 Find the value of x in each of these diagrams.
They are not drawn accurately.

(a)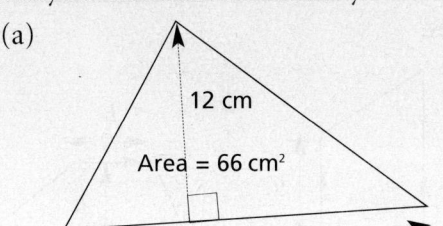

12 cm

Area = 66 cm²

x

(b)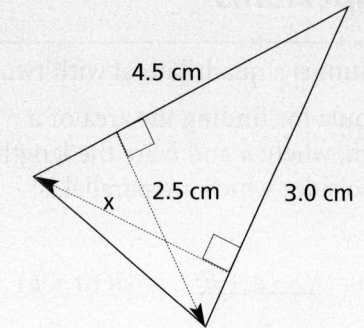

4.5 cm

2.5 cm   3.0 cm

x

B6 Write, as simply as possible, an expression for the area of each of these shapes.

(a)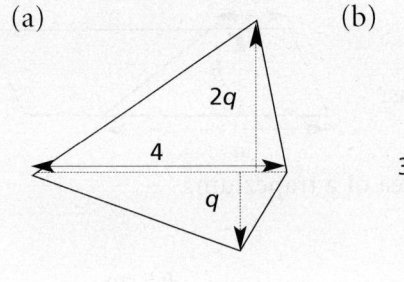

2q

4

q

(b)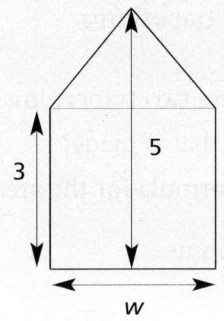

5

3

w

(c)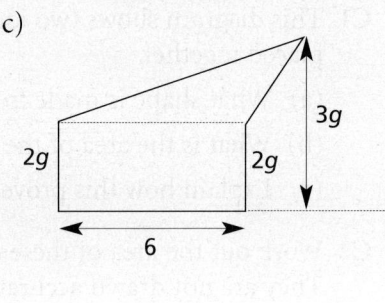

3g

2g   2g

6

*__B7__ (a) Write down the area of these triangles.

(i)

2.8 m

4.5 m

(ii)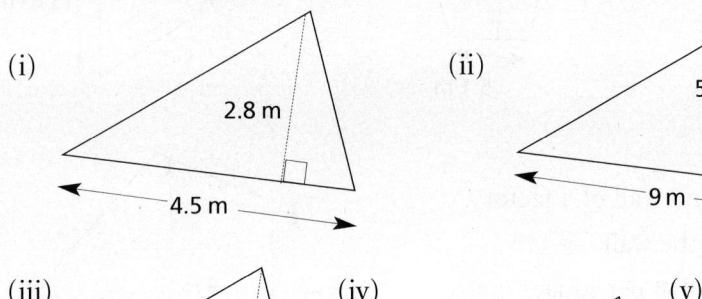

5.6 m

9 m

(iii)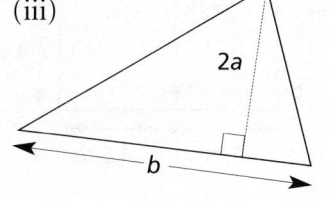

2a

b

(iv)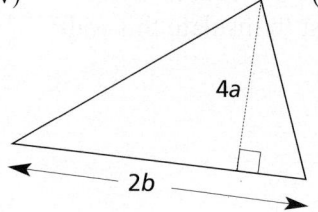

4a

2b

(v)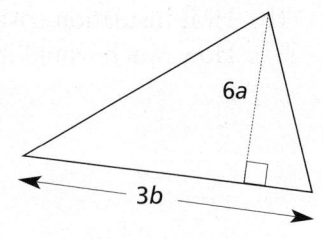

6a

3b

(b) What happens to the area of a triangle when its height and base are doubled?

(c) What happens to the area of a triangle when it is enlarged by scale factor 3?

C **Trapeziums**

A trapezium is a quadrilateral with two parallel sides.

The formula for finding the area of a trapezium, where a and b are the lengths of the two sides which are parallel is:

$$A = \frac{h \times (a + b)}{2} = \frac{1}{2}h\,(a + b)$$

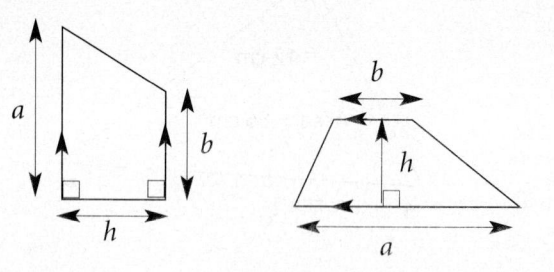

C1 This diagram shows two identical trapeziums placed together.

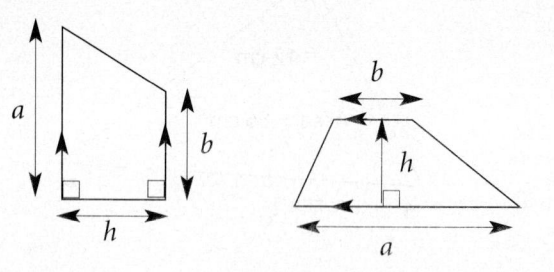

(a) What shape is made from these two trapeziums?

(b) What is the area of the shape that is made?

(c) Explain how this proves the formula for the area of a trapezium.

C2 Work out the area of these trapeziums. They are not drawn accurately.

(a)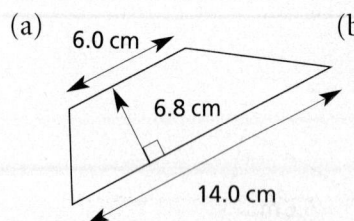

6.0 cm

6.8 cm

14.0 cm

(b)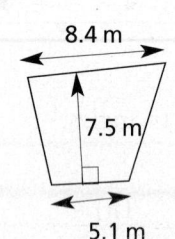

8.4 m

7.5 m

5.1 m

(c)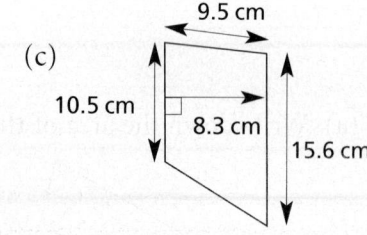

9.5 cm

10.5 cm

8.3 cm

15.6 cm

C3 This diagram shows the end wall of a factory.

(a) Work out the area of the wall.

(b) Heat insulation costs £18 per square metre. How much would it cost to insulate this wall?

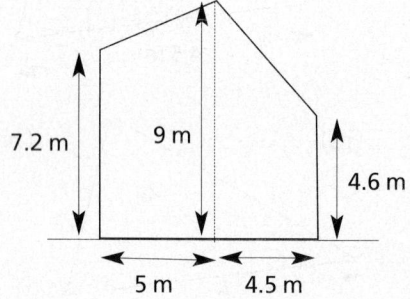

7.2 m

9 m

4.6 m

5 m

4.5 m

C4 This diagram shows the keel of a yacht. Find the area of this keel?

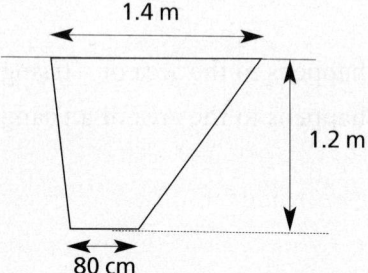

1.4 m

1.2 m

80 cm

C5 Find the missing lengths in these trapeziums.

(a)

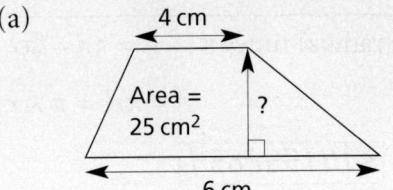

4 cm

Area = 25 cm²

?

6 cm

(b)

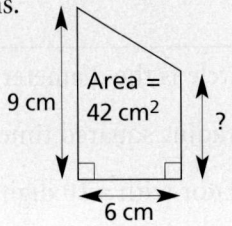

9 cm

Area = 42 cm²

?

6 cm

*C6 Find the missing lengths in these trapeziums.

(a)

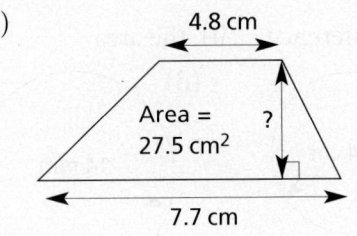

4.8 cm

Area = 27.5 cm²

?

7.7 cm

(b)

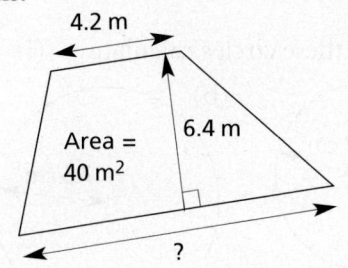

4.2 m

Area = 40 m²

6.4 m

?

Areas under curves

Trapeziums can be used to find approximations to 'areas under curves' where the other sides are straight lines such as axes on a graph

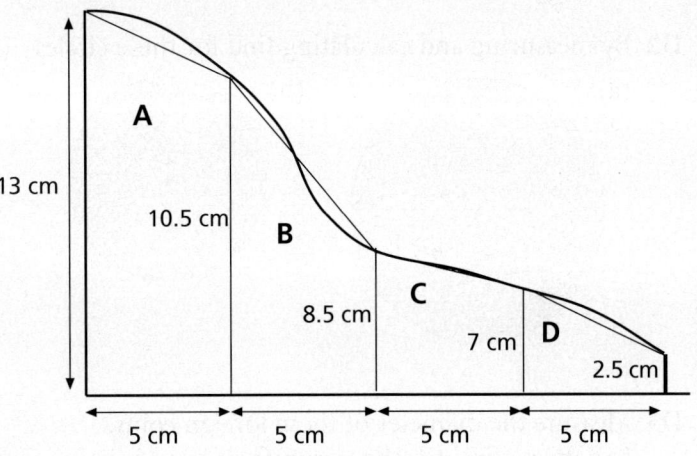

13 cm

10.5 cm

A

B

8.5 cm

C

7 cm

D

2.5 cm

5 cm   5 cm   5 cm   5 cm

The approximate area under this curve can be found by finding the area of trapeziums A, B, C, and D.

As all the trapeziums are given width 5 cm:

Area = $[\frac{1}{2}(13 + 10.5) \times 5] + [\frac{1}{2}(10.5 + 8.5) \times 5] + [\frac{1}{2}(8.5 + 7) \times 5] + [\frac{1}{2}(7 + 2.5) \times 5]$

or more simply

$= \frac{1}{2} \times 5 \times [13 + 10.5 + 10.5 + 8.5 + 8.5 + 7 + 7 + 2.5] = 168.75 \text{ cm}^2$

How good an approximation do you think this is?

> The **trapezium rule** for an approximation to the area under a curve is:
> Area = $\frac{1}{2}$(width of strip) × (1st height + (2 × each middle height) + last height)

Find the areas under the curves on sheet P12.

- make all your trapeziums 2 cm wide
- round all the heights to the nearest half centimetre.

D Circles

The circumference of a circle is the diameter (twice the radius) times π. $C = \pi d = 2\pi r$

The area of a circle is the radius squared times π. $A = \pi r^2 = \pi \times r \times r$

The value of π on a calculator with a 10 digit display is $\boxed{3.141592654}$

In these questions use the π button on your calculator.
Round your answers to 1 decimal place.

D1 For each of these circles calculate   (i) the circumference   (ii) the area

(a) 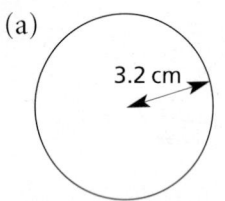 3.2 cm

(b) 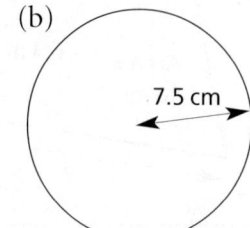 7.5 cm

(c) 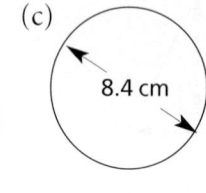 8.4 cm

(d) 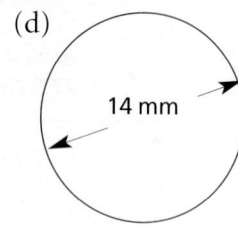 14 mm

D2 By measuring and calculating find for these circles   (i) the circumference   (ii) the area.

(a)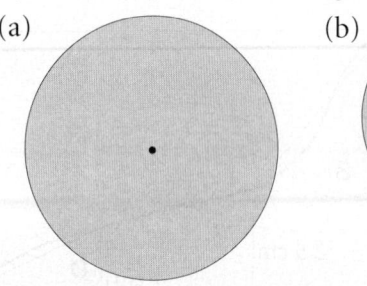

(b)

(c)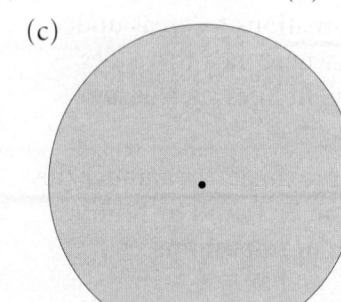

D3 Measure the diameter of these foreign coins.
Use this to find (i) the circumference and (ii) area of each one.

(a) Greek
100 drachma

(b) Irish 1p

(c) Tunisian
$\frac{1}{2}$ dinar

(d) Hong Kong
10 cents

D4 The Millennium Dome at Greenwich is 390 m in diameter.

    (a) How long is it round the edge of the Dome to the nearest metre?

    (b) What is the area of ground floor space inside the dome?
        (Give your answer to the nearest square metre.)

D5 This running track has semicircles at each end

    (a) How much further would someone running
        around the outside of the track run compared
        with someone running around the inside?
        (Give your answer to the nearest metre)

    (b) What is the area of the track (shaded)?
        (Give your answer to the nearest square metre.)

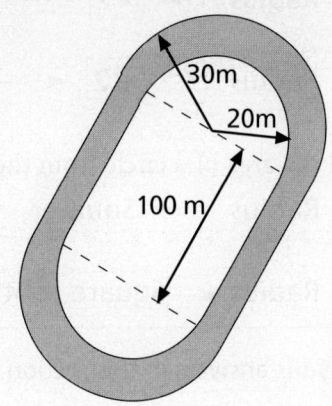

D6 A company cuts bottle tops 25 mm in diameter from
a piece of metal foil as shown in pattern A.
A designer suggests that there would be less wastage if the tops
were cut from a different sized piece of metal as shown in B.

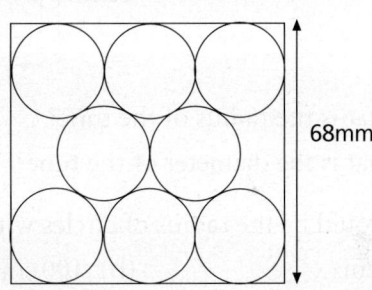

    (a) What is the area of foil used for the bottle tops?

    (b) Find the area of the rectangle in each case.

    (c) What percentage of foil is used in each case?

    (d) Was the designer right?

D7 (a) What is the area of the shaded
        circle in this diagram?

    (b) What percentage of the area of
        the square is the shaded circle?

    (c) Use trial and improvement to
        find to 2d.p, the radius, of a circle which
        would be half the area of this square.

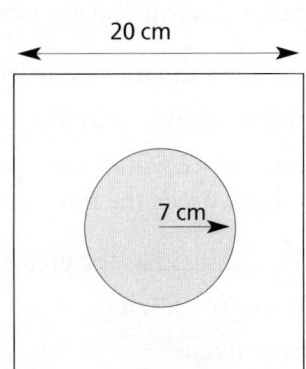

E *Finding the radius*

To find the radius of a circle from the circumference the operations are easily reversed:

Radius $\longrightarrow$ $\boxed{\times 2}$ $\longrightarrow$ $\boxed{\times \pi}$ $\rightarrow$ Circumference

Radius $\longleftarrow$ $\boxed{\div 2}$ $\longleftarrow$ $\boxed{\div \pi}$ $\longleftarrow$ Circumference

To find the area of a circle from the radius these are the operations:

Radius $\longrightarrow$ $\boxed{\text{Square}}$ $\longrightarrow$ $\boxed{\times \pi}$ $\rightarrow$ Area

Radius $\longleftarrow$ $\boxed{\text{Square root}}$ $\longleftarrow$ $\boxed{\div \pi}$ $\longleftarrow$ Area

Give all your answers in this section to 1 decimal place.

E1 A bicycle wheel has a circumference of 200 cm. What is the radius of the wheel?

E2 A rectangular piece of metal is curved to make a tube.

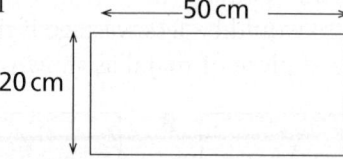 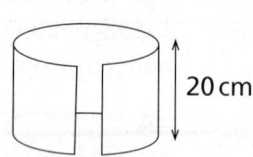

50 cm

20 cm   20 cm

    (a) What is the radius of the tube?

    (b) What is the diameter of the tube?

E3 What would be the radius of circles with areas:

    (a) $50\,\text{cm}^2$ (b) $100\,\text{cm}^2$ (c) $25\,\text{m}^2$

E4 The roughly circular crater Copernicus on the moon has an area of $3200\,\text{km}^2$. What is the diameter of the crater to the nearest kilometre?

E5 (a) A farmer wants to make a circular pen which contains an area of ground $200\,\text{m}^2$. What would be the radius of the circle needed?

    (b) How long a piece of fencing would she need to go round this circle?

E6 A piece of land is a square of side length 50 m.

    (a) What radius would a circular piece of land need in order to cover the same area?

    (b) How much fence would the square and circular fields need around the edge? Which shape needs the most fencing?

E7 Which of these circles has the greatest radius?

    Circle A: Diameter 12.5 cm   Circle B: Circumference 40 cm

    Circle C: Area $105\,\text{cm}^2$

F *A mixed bag*

F1 (a) Which of these shapes do you think has the largest area?
Find the area of each shape and check your estimation.

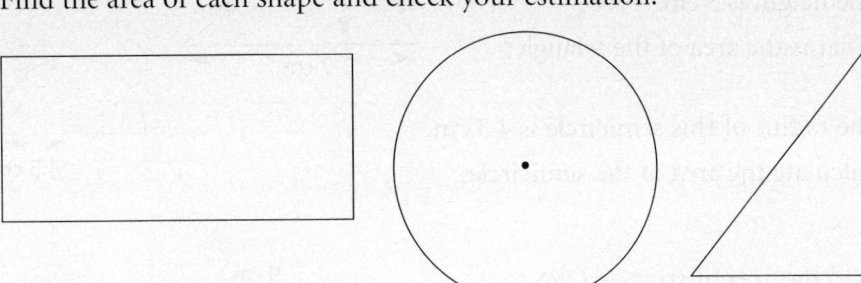

(b) Work out the perimeter of each shape.
Which shape has the largest perimeter?

F2 A farmer has a length of fence 240 m long.
How big an area would this surround if it was set out as

(a) a square

(b) a rectangle which is twice as long as it is wide

(c) a circle

(d) a right angled triangle with sides 60 m, 80 m and 100 m long?

F3 The circle in this pattern has radius 2.3 cm.
Find the area of

(a) one of the grey shaded triangles

(b) the square

(c) one of the black segments

F4 These diagrams have been drawn on centimetre square paper.
For each one work out which of the two shapes has the larger area.

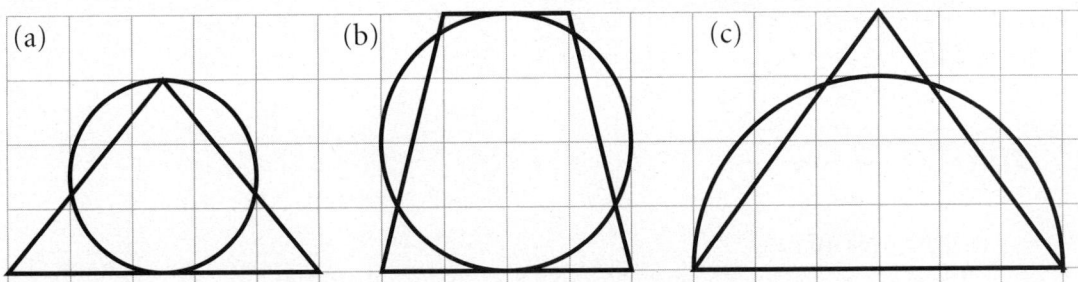

Test yourself with these questions

T1 (a) The base of this triangle is 7 cm.
The height is 5 cm.

What is the area of the triangle?

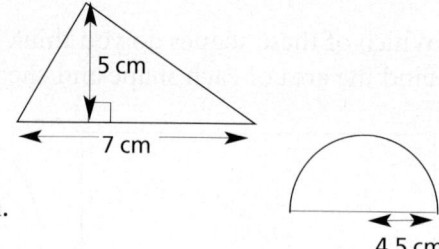

(b) The radius of this semicircle is 4.5 cm.

Calculate the area of the semicircle.

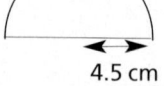

4.5 cm

T2 (a) Find the area of triangle QRS.

(b) Find the area of trapezium PQRS.
Remember to state the units in
both your answers.

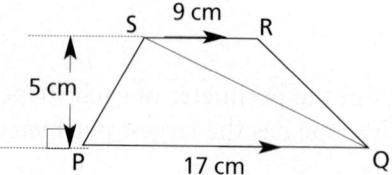

T3 A large clock-face on a building is a circle with radius 2.91 metres.

(a) The minute hand reaches to the perimeter of the clock face.
How far does the tip of a minute hand travel in an hour?

(b) What is the area of the clock face?

T4 The distance round a circular running track is 1000 m.

What is its diameter?

T5 This diagram shows two shapes,
a parallelogram and a trapezium.

They both have the same area.

Find the height of the parallelogram,
marked h on the diagram.

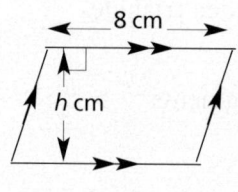

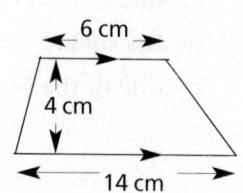

9 Fractions 1

You should know

- what is meant by equivalent fractions
- how to put decimals, for example 0.7, 0.62, 0.09, 0.115, in order of size
- how to rearrange simple formulas

You will learn

- how to put fractions in order of size
- how to work with mixed numbers, for example $4\frac{2}{3}$
- how to add and subtract fractions
- how to express one quantity as a fraction of another
- how to change decimals to fractions and fractions to decimals

A Equivalent fractions

 This unit is intended to be done without a calculator.

A1 Use these diagrams to help you write down three fractions equivalent to $\frac{2}{3}$.

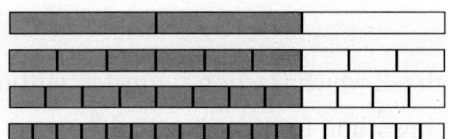

To find a fraction equivalent to $\frac{2}{3}$,
multiply numerator (top) and denominator (bottom) by the same number.

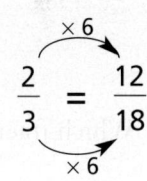

$$\frac{2}{3} = \frac{12}{18}$$

A2 (a) What is the multiplier here? (b) Complete the second fraction.

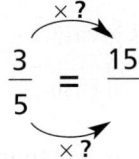

$$\frac{3}{5} = \frac{15}{}$$

A3 Copy these and find the missing numbers.

(a) $\frac{2}{3} = \frac{10}{}$ (b) $\frac{5}{6} = \frac{20}{}$ (c) $\frac{3}{4} = \frac{9}{}$ (d) $\frac{2}{5} = \frac{12}{}$ (e) $\frac{3}{7} = \frac{21}{}$

A4 Copy these and find the missing numbers.

(a) $\frac{3}{8} = \frac{}{32}$ (b) $\frac{5}{7} = \frac{}{35}$ (c) $\frac{4}{9} = \frac{}{36}$ (d) $\frac{5}{8} = \frac{}{24}$ (e) $\frac{7}{12} = \frac{}{60}$

Cancelling

Sometimes a fraction can be **simplified** by dividing numerator and denominator by the same number.

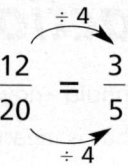

This is called **cancelling** the **common factor** 4.

If the fraction can't be simplified any more, it is in its **lowest terms**, or **simplest form**.

A5 Write each of these fractions in its lowest terms.

(a) $\frac{20}{30}$ (b) $\frac{3}{9}$ (c) $\frac{12}{16}$ (d) $\frac{20}{25}$ (e) $\frac{30}{36}$

(f) $\frac{21}{28}$ (g) $\frac{36}{60}$ (h) $\frac{26}{39}$ (i) $\frac{22}{55}$ (j) $\frac{72}{96}$

A6 Sort these into pairs of equivalent fractions.

$\frac{7}{28}$ $\frac{5}{8}$ $\frac{2}{5}$ $\frac{3}{4}$ $\frac{20}{24}$ $\frac{12}{16}$ $\frac{1}{4}$ $\frac{9}{21}$ $\frac{3}{7}$ $\frac{15}{24}$ $\frac{8}{20}$ $\frac{5}{6}$

B *Ordering fractions*

It is easy to see from these diagrams that $\frac{2}{3}$ is greater than $\frac{7}{12}$.

We can tell which fraction is greater, **without** diagrams, by changing $\frac{2}{3}$ into twelfths:

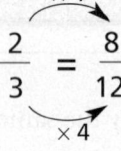

B1 Which fraction is greater, $\frac{3}{4}$ or $\frac{17}{20}$? (Change $\frac{3}{4}$ to twentieths.)

B2 Which fraction is greater, $\frac{4}{5}$ or $\frac{11}{15}$?

B3 Work out which fraction in each pair is greater.

(a) $\frac{1}{3}, \frac{5}{12}$ (b) $\frac{11}{16}, \frac{3}{4}$ (c) $\frac{19}{30}, \frac{7}{10}$ (d) $\frac{2}{5}, \frac{7}{20}$ (e) $\frac{3}{8}, \frac{11}{24}$

B4 Pat wants to know whether $\frac{3}{5}$ is greater or less than $\frac{2}{3}$.

He makes a list of fractions equivalent to $\frac{3}{5}$, ...

$$\frac{3}{5} = \frac{6}{10} = \frac{9}{15} = \frac{12}{20} = \frac{15}{25} =$$

and a list of fractions equivalent to $\frac{2}{3}$.

$$\frac{2}{3} = \frac{4}{6} = \frac{6}{9} = \frac{8}{12} = \frac{10}{15} =$$

How can you tell from the lists whether $\frac{3}{5}$ is greater or less than $\frac{2}{3}$?

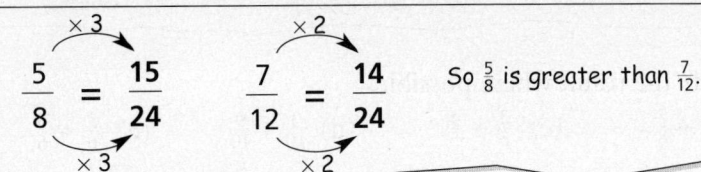
B5 Work out which fraction is greater, $\frac{3}{8}$ or $\frac{2}{5}$.

B6 Work out which fraction in each pair is greater.

(a) $\frac{1}{3}$, $\frac{2}{5}$ (b) $\frac{3}{5}$, $\frac{5}{8}$ (c) $\frac{5}{6}$, $\frac{7}{8}$ (d) $\frac{7}{8}$, $\frac{4}{5}$ (e) $\frac{5}{8}$, $\frac{7}{10}$

B7 Work out which fraction in each pair is greater.

(a) $\frac{5}{9}$, $\frac{3}{5}$ (b) $\frac{4}{7}$, $\frac{5}{8}$ (c) $\frac{3}{10}$, $\frac{1}{3}$ (d) $\frac{2}{5}$, $\frac{3}{7}$ (e) $\frac{3}{8}$, $\frac{5}{12}$

Ⓒ *Mixed numbers*

A **mixed number** is made up of a whole number and a fraction, for example $2\frac{3}{4}$.

A mixed number can be changed to an **improper** ('top heavy') fraction.

For example, $2\frac{3}{4} = 1 + 1 + \frac{3}{4} = \frac{4}{4} + \frac{4}{4} + \frac{3}{4} = \frac{11}{4}$.

C1 Change these mixed numbers to improper fractions.

(a) $1\frac{1}{4}$ (b) $2\frac{1}{3}$ (c) $1\frac{2}{3}$ (d) $4\frac{1}{2}$ (e) $3\frac{2}{5}$

(f) $2\frac{2}{3}$ (g) $3\frac{1}{5}$ (h) $1\frac{5}{8}$ (i) $2\frac{1}{10}$ (j) $4\frac{7}{8}$

 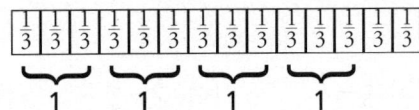 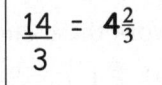
C2 Change these improper fractions to mixed numbers.

(a) $\frac{7}{2}$ (b) $\frac{13}{4}$ (c) $\frac{13}{5}$ (d) $\frac{11}{6}$ (e) $\frac{15}{4}$

C3 Change these improper fractions to mixed numbers.

(a) $\frac{22}{4}$ (b) $\frac{13}{6}$ (c) $\frac{20}{7}$ (d) $\frac{14}{5}$ (e) $\frac{29}{10}$

Ⓓ *Addition and subtraction*

It is easy to add and subtract fractions with the same denominator.

$$\frac{3}{5} + \frac{1}{5} = \frac{4}{5} \qquad \frac{5}{8} + \frac{7}{8} = \frac{12}{8} = 1\frac{4}{8} = 1\frac{1}{2} \qquad 1\frac{1}{3} - \frac{2}{3} = \frac{4}{3} - \frac{2}{3} = \frac{2}{3}$$

D1 Work these out. Simplify the result where possible.

   (a) $\frac{2}{7} + \frac{3}{7}$ (b) $\frac{1}{8} + \frac{5}{8}$ (c) $\frac{3}{5} + \frac{4}{5}$ (d) $\frac{3}{10} + \frac{9}{10}$ (e) $\frac{5}{6} + \frac{5}{6}$

D2 Work these out. Simplify the result where possible.

   (a) $1\frac{1}{5} + \frac{3}{5}$ (b) $2\frac{3}{4} + \frac{3}{4}$ (c) $1\frac{2}{3} + \frac{2}{3}$ (d) $\frac{5}{6} + 1\frac{1}{6}$ (e) $1\frac{3}{8} + 1\frac{1}{8}$

D3 Work out these subtractions. Simplify where possible.

   (a) $\frac{4}{5} - \frac{1}{5}$ (b) $\frac{7}{8} - \frac{3}{8}$ (c) $\frac{9}{10} - \frac{3}{10}$ (d) $1\frac{5}{8} - \frac{1}{8}$ (e) $2\frac{1}{4} - \frac{3}{4}$

Worked example

Work out $\frac{3}{4} + \frac{1}{6}$.

We need to change $\frac{3}{4}$ and $\frac{1}{6}$ into equivalent fractions with the **same denominator**.
The denominator must be a multiple of 4 and also a multiple of 6. So **12** will be OK.

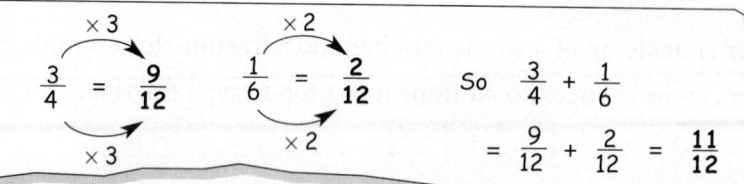

$$\frac{3}{4} = \frac{9}{12} \qquad \frac{1}{6} = \frac{2}{12} \qquad \text{So} \quad \frac{3}{4} + \frac{1}{6}$$

$$= \frac{9}{12} + \frac{2}{12} = \frac{11}{12}$$

D4 Work these out.

   (a) $\frac{1}{4} + \frac{1}{3}$ (b) $\frac{2}{3} + \frac{1}{4}$ (c) $\frac{1}{6} + \frac{1}{4}$ (d) $\frac{1}{5} + \frac{2}{3}$ (e) $\frac{2}{5} + \frac{1}{4}$

D5 Work these out.

   (a) $\frac{2}{5} - \frac{1}{4}$ (b) $\frac{3}{4} - \frac{1}{6}$ (c) $\frac{2}{3} - \frac{1}{4}$ (d) $\frac{3}{4} - \frac{2}{3}$ (e) $\frac{1}{2} - \frac{2}{5}$

D6 Work these out.

   (a) $\frac{2}{5} + \frac{1}{3}$ (b) $\frac{2}{5} - \frac{1}{4}$ (c) $\frac{3}{4} - \frac{2}{5}$ (d) $\frac{1}{8} + \frac{5}{6}$ (e) $\frac{1}{12} + \frac{2}{5}$

D7 Work these out. The results will be mixed numbers.

   (a) $\frac{3}{8} + \frac{2}{3}$ (b) $\frac{5}{6} + \frac{1}{4}$ (c) $\frac{2}{3} + \frac{2}{5}$ (d) $\frac{5}{8} + \frac{2}{3}$ (e) $1\frac{3}{4} + \frac{2}{5}$

D8 Work these out.

   (a) $\frac{3}{8} - \frac{1}{5}$ (b) $\frac{5}{8} + \frac{1}{3}$ (c) $1\frac{2}{3} + \frac{5}{6}$ (d) $1\frac{1}{3} - \frac{1}{8}$ (e) $\frac{7}{8} + \frac{2}{3}$

   (f) $\frac{4}{5} - \frac{1}{8}$ (g) $\frac{7}{8} - \frac{1}{3}$ (h) $\frac{5}{6} + \frac{1}{8}$ (i) $\frac{7}{8} - \frac{5}{6}$ (j) $1\frac{2}{5} - \frac{2}{3}$

E Expressing one quantity as a fraction of another

> **Worked example**
>
> In a class of 36 children, 15 were absent with flu.
> What fraction of the class was absent?
>
> Each child is $\frac{1}{36}$ of the class.
> So $\frac{15}{36}$ of the class were absent.
>
> This simplifies to $\frac{5}{12}$ (by dividing top and bottom by 3).
>
> 15   36

E1 In a typical 24-hour period, Karl spends 6 hours working, 8 hours sleeping,
3 hours eating and the rest of the time doing other things.

Write, in its simplest form, the fraction of the time Karl spends

(a) working   (b) sleeping   (c) eating   (d) doing other things

E2 Write each of these as a fraction in its simplest form.

(a) 12 out of 20   (b) 8 out of 30   (c) 12 out of 16   (d) 25 out of 40

E3 Gert owns 30 acres of land: 12 acres are woodland, 15 are grass and the rest is marsh.
Write, in its simplest form, the fraction of Gert's land that is

(a) woodland   (b) grass   (c) marsh

E4 There are 15 boys and 12 girls in a class.
What fraction of the class (in its simplest form) are boys?

E5 Last season Brockleton United won 18 matches, drew 8 and lost 10.
What fraction, in its simplest form, of the matches were

(a) won   (b) drawn   (c) lost

E6 Write each of these fractions in its simplest form:

(a) the fraction of the circle that is red

(b) the fraction that is blue

(c) the fraction that is green

(d) the fraction that is yellow

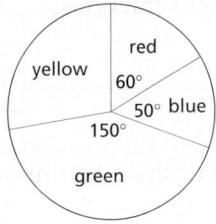

E7 Joe had £4 pocket money. He spent £1.50 on sweets.
What fraction of his pocket money did he spend on sweets?

E8 Write each of these as a fraction in its lowest terms.

(a) £2.00 out of £2.50   (b) £2.50 out of £7.50   (c) £2.40 out of £3.00

F Changing between fractions and decimals

Worked examples

Change each of these decimals to a fraction, in its lowest terms: (a) 0.32   (b) 0.275

(a) 0.32 means the same as $\frac{32}{100}$.

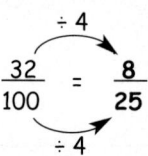

$$\frac{32}{100} = \frac{8}{25}$$

(b) 0.275 means the same as $\frac{275}{1000}$.

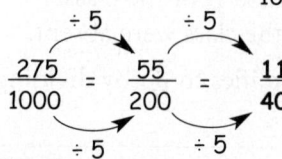

$$\frac{275}{1000} = \frac{55}{200} = \frac{11}{40}$$

F1 Change each of these decimals to a fraction, in its lowest terms.

(a) 0.6   (b) 0.64   (c) 0.625   (d) 0.35   (e) 0.08

(f) 0.825   (g) 0.035   (h) 0.72   (i) 0.004   (j) 0.152

Worked example

Change $\frac{3}{8}$ to a decimal.

You need to divide 3 by 8.
To allow for decimal places, write 3 as 3.00…

$$\begin{array}{r} 0.375 \\ \hline 8)3.0^60^40 \end{array} \qquad \frac{3}{8} = 0.375$$

F2 Change each of these fractions to a decimal.

(a) $\frac{1}{8}$ (b) $\frac{5}{8}$ (c) $\frac{7}{8}$ (d) $\frac{1}{16}$ (e) $\frac{5}{16}$

Fractions with denominator 50, 20 or 25 can be changed to decimals
through equivalent fractions.

$$\frac{9}{50} = \frac{18}{100} = 0.18 \qquad \frac{7}{20} = \frac{35}{100} = 0.35 \qquad \frac{6}{25} = \frac{24}{100} = 0.24$$

F3 Write each of these lists in order, smallest first.

(a) 0.6, $\frac{5}{8}$, $\frac{13}{20}$, 0.59   (b) $\frac{37}{50}$, 0.7, $\frac{4}{5}$, 0.77   (c) $\frac{3}{10}$, 0.35, $\frac{9}{20}$, 0.4

(d) 0.25, $\frac{3}{8}$, 0.3, $\frac{7}{20}$ (e) 0.4, 0.405, $\frac{9}{20}$, 0.5   (f) $\frac{7}{8}$, $\frac{3}{4}$, 0.85, 0.8

F4 Decode this message. Rewrite the letters in the order of the numbers, smallest first.

O	F	T	E	N	T	O	S	I	L	E	N	C	E
0.3	$\frac{9}{20}$	$\frac{1}{2}$	$\frac{1}{10}$	$\frac{3}{8}$	$\frac{1}{5}$	0.085	0.12	$\frac{1}{4}$	$\frac{2}{5}$	0.408	0.09	0.13	$\frac{1}{8}$

⒢ Recurring decimals

$$\frac{1}{3} = ? \qquad \frac{1}{6} = ? \qquad \frac{1}{11} = ? \qquad \frac{1}{7} = ?$$

G1 Change these to decimals: (a) $\frac{2}{3}$ (b) $\frac{1}{9}$ (c) $\frac{2}{9}$ (d) $\frac{4}{9}$ (e) $\frac{7}{9}$

G2 Change each of these fractions to recurring decimals.
What do you notice about the results?
(a) $\frac{1}{7}$ (b) $\frac{2}{7}$ (c) $\frac{3}{7}$ (d) $\frac{4}{7}$ (e) $\frac{5}{7}$ (f) $\frac{6}{7}$

G3 Change each of these fractions to decimals.
(a) $\frac{5}{6}$ (b) $\frac{2}{11}$ (c) $\frac{3}{11}$ (d) $\frac{1}{12}$ (e) $\frac{1}{13}$

Investigation	How can you tell when a fraction will give a recurring decimal? Which of these will be recurring: $\frac{1}{15}$ $\frac{1}{16}$ $\frac{1}{17}$ $\frac{1}{30}$ $\frac{1}{40}$

Test yourself with these questions

T1 Find the missing numbers a, b, c and d.

$$\frac{2}{5} = \frac{12}{a} \qquad \frac{3}{8} = \frac{b}{40} \qquad \frac{5}{c} = \frac{20}{36} \qquad \frac{d}{4} = \frac{24}{32}$$

T2 Write each list in order, starting with the smallest.
(a) 0.04, $\frac{3}{20}$, 0.1, $\frac{1}{50}$ (b) $\frac{7}{8}$, 0.9, $\frac{17}{20}$, 0.86

T3 Work these out.
(a) $\frac{1}{5} + \frac{2}{3}$ (b) $\frac{2}{3} - \frac{1}{4}$ (c) $\frac{3}{8} + \frac{4}{5}$ (d) $3\frac{1}{4} - \frac{5}{8}$

T4 Fabia planted 80 lettuce seeds but only 25 of them grew into plants.
What fraction of the seeds grew? Write it in its lowest terms.

T5 A landowner died. In his will the land was to be shared between his three children.
The eldest inherited $\frac{2}{5}$ of the land and the second child inherited $\frac{1}{3}$.
What fraction did the youngest inherit?

T6 On a computer keyboard there are 104 keys.
26 of the keys have letters on them.

What fraction of the keys have letters on them?
(Give your answer in its simplest form.)   [AQA (NEAB) 1998]

T7 (a) Change $\frac{3}{8}$ to a decimal.
(b) Use your answer to part (a) to write $\frac{3}{80}$ as a decimal.   [OCR]

10 Substitution

What you should know ...

◆ How to work out expressions such as $\frac{h-6}{10}$, $35 - 2h$, $4h^2 - 40$, $\frac{h^2}{10}$ and $200 - h^2$ when you know the value of h.

A Review

A1 In your head, work out the value of each of these expressions when $h = 4$.

    (a) $2h^2$ (b) $100 - h^2$ (c) $\frac{100}{5h}$ (d) $\frac{64}{h^2}$ (e) $\frac{1 - h^2}{5}$

A2 Use a calculator to work out:

    (a) $8a^2$ when $a = 2.5$ (b) $7.2 - 2b^2$ when $b = 1.1$ (c) $4.5c - 2.7$ when $c = 1.5$

    (d) $\frac{9}{2d - 3}$ when $d = 2.4$ (e) $1.5(2.2e + 1)$ when $e = 2$ (f) $\frac{f^2}{5} - 3f$ when $f = 4.5$

A3 The circumference of a circle is given by the formula $C = 2\pi r$.
C stands for the circumference and r for the radius.

Work out the circumference of each of these, in centimetres.
Give your answers to one decimal place.

    (a) 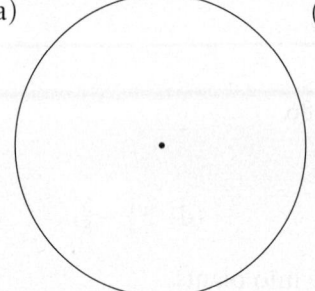   (b) 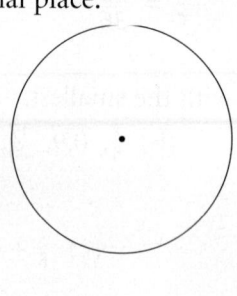   (c)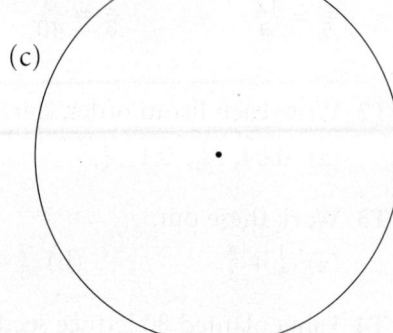

A4 The area of a circle is given by the formula $A = \pi r^2$.
Work out the areas of the circles in question A3.
Give each answer to the nearest $0.1\,\text{cm}^2$.

A5 (a) Which of the expressions in the boxes has the highest value when $x = 5$?

                          $2 - 4x^2$ $10(x + 3)$ $2(10 - x)$

    (b) Which of the expressions has the lowest value when $x = 5$?

                                 $\frac{x^2}{2}$ $\frac{25}{x - 4}$

A6 Which numbers in this list fit each equation below?
More than one number may fit an equation.

                                        $^-5$ $^-1$ 1 3.5 4 5

    (a) $5x - 20 = 0$ (b) $2x^2 - 50 = 0$ (c) $4(2x - 5) = 8$ (d) $\frac{1}{x^2} = 1$

B More letters

When you have several letters in an expression, you treat them just like numbers.

Evaluate $a - bc$ when $a = 5, b = 2, c = 8$.

$a - bc$

bc means $b \times c$

$= 5 - 2 \times 8$

$= 5 - 16 = {}^-11$

Evaluate $a(b - c)^2$ when $a = 5, b = 2, c = 8$

$a(b - c)^2$

$a(b - c)^2$ means $a \times (b - c)^2$

$= 5 \times (2 - 8)^2$

$= 5 \times ({}^-6)^2 \ = 5 \times 36 \ = 180$

B1 Calculate in your head the values of each of the following
when $p = 2$, $q = 3$, $r = 6$

(a) $r(p + q)$ (b) $rp + q$ (c) $\dfrac{r}{pq}$ (d) $\dfrac{q + r}{p}$ (e) $5r^2$

Check that you can make your calculator agree with you.

B2 Calculate the values of each of the following in your head
when $f = 2$, $g = {}^-6$, $h = 4$

(a) $f + gh$ (b) $\dfrac{f}{g - h}$ (c) $(f + g)^2$ (d) $10g^2$ (e) $f^2 + h^2$

Check that you can make your calculator agree with you.

B3 Evaluate each of the expressions in question B2
when $f = 7.9$, $g = 5.7$, $h = 3.8$
Give your answers accurate to 1 decimal place.

B4 If $u = 2.1$, $v = 3.1$ and $w = 5.9$, evaluate each of these expressions
giving answers accurate to 2 decimal places.

(a) $\dfrac{u}{v + w}$ (b) $w - \dfrac{u}{v}$ (c) $u + vw$ (d) $\dfrac{u + v}{v + w}$ (e) uv^2

B5 Evaluate each of these expressions when
$a = {}^-4.5$, $b = 0.5$ and $c = {}^-2.5$.

(a) $\dfrac{a^2}{b^2 + c^2}$ (b) $(c - 2b)^2$ (c) $3b^2$ (d) $(3b)^2$ (e) $ab + bc + ca$

B6 Evaluate each of the following expressions when
$w = \frac{1}{2}$, $x = \frac{1}{3}$ and $y = \frac{3}{4}$.

(a) $2w$ (b) $2w - y$ (c) $w + x$ (d) $w + x + y$ (e) $w + x - y$

B7 If $a = 2$ and $b = 3$, calculate in your head the values of

(a) a^2b (b) ab^2 (c) $(ab)^2$

If any of your answers are the same, you have made a mistake. In that case, find it!

C Units

When using a formula for a real problem, you must be careful about the units.

To find the area of this triangle we must work entirely in either centimetres or millimetres.

The area of a triangle is given by $A = \frac{1}{2}bh$ or $\frac{bh}{2}$ where b is the base and h is the height.

Working in centimetres, $A = \frac{bh}{2} = \frac{2.5 \times 1.8}{2}$

$$= \frac{4.5}{2} = 2.25$$

So the area is 2.25 cm^2.

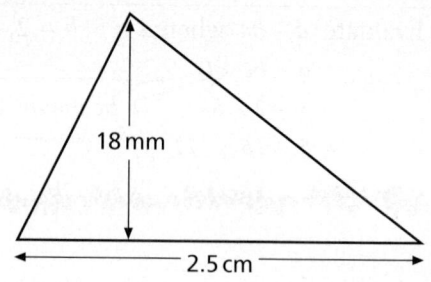

18 mm

2.5 cm

When using formulas

- it must be clear what each letter stands for
- think carefully about the units of any quantities
- give any answers complete with units

C1 The area of a triangle is given by the formula $A = \frac{bh}{2}$

Work out the areas of triangles where

(a) $b = 25\text{ cm}, \ h = 10\text{ cm}$

(b) $b = 1.2\text{ m}, \ h = 90\text{ cm}$

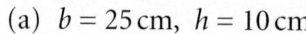

h

b

C2 The area, A square units, of a trapezium is given by the formula $A = \frac{1}{2}(a + b)h$. (a, b and h are shown on the diagram.)

Use the formula to calculate the areas of trapeziums for which:

(a) $a = 60\text{ cm}, \ b = 80\text{ cm}, \ h = 40\text{ cm}$

(b) $a = 50\text{ cm}, \ b = 1.8\text{ m}, \ h = 80\text{ cm}$.

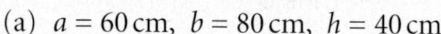

a

h

b

C3 The power, W, used by an electrical appliance is measured in watts.
The current, A, the appliance needs in amps is given by the formula $A = \frac{W}{220}$

(a) Work out the current needed by:

   (i) A 250 watt television   (ii) An 80 watt computer

   (iii) A 60 watt light bulb   (iv) A 7 kilowatt cooker 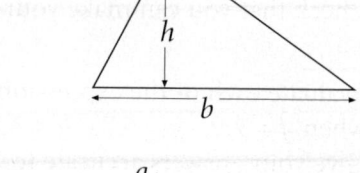 *1 kilowatt = 1000 watts*

   (Give your answers to 1 significant figure.)

(b) A shower needs a current of 30 amps.
How many kilowatts does it use?

C4 Calculate the area of each of these trapeziums.
Make sure you include units in your answers.

(a)

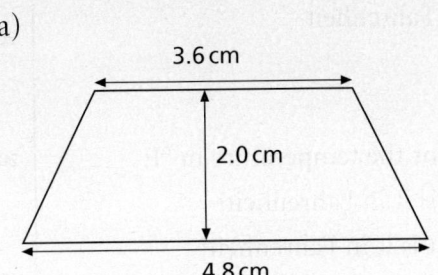

(b)

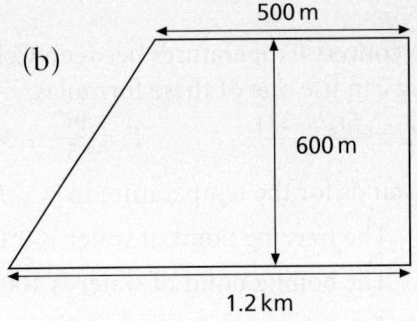

C5 A rectangle has length L and width W.
Its perimeter, P (the total distance round the edge),
can be found using the formula: $P = 2(L + W)$

Use this formula to find the perimeters
of the following rectangles:

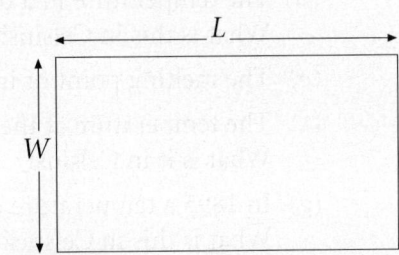

(a) a field 120 m long and 200 m across

(b) a poster 40 cm wide and 65 cm high

(c) a postage stamp measuring 18 mm across and 2.5 cm from top to bottom

(d) a doormat measuring 1.2 m by 75 cm.

C6 This cuboid has square ends.
Its volume, V, is given by the formula $V = ab^2$.

Calculate the volumes of each of these cuboids
with square ends.

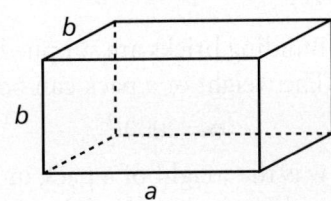

(a)

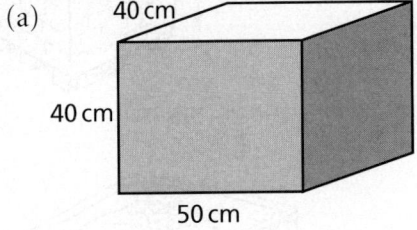

(b)

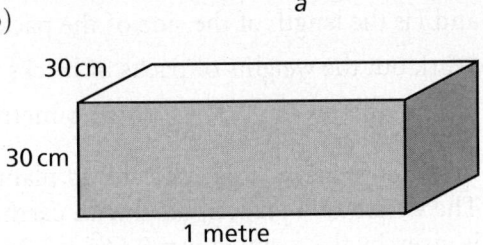

***C7** A cuboid with square ends, as in question C6,
has a surface area, A, given by the formula $A = 4ab + 2b^2$.

Calculate the surface area of

(a) a cuboid that is 10 cm long, with ends 4 cm by 4 cm

(b) a cuboid that is 1.2 metres long, with ends 75 centimetres square.

D Mixed examples

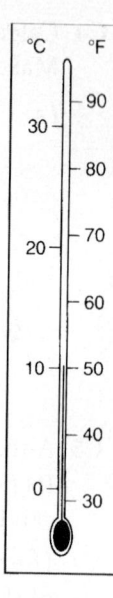

D1 To convert temperatures between Celsius and Fahrenheit you can use one of these formulas:

$$C = \frac{5(F - 32)}{9} \qquad F = \frac{9C}{5} + 32$$

C stands for the temperature in °C, F stands for the temperature in °F.

(a) The freezing point of water is 0°C. What is it in Fahrenheit?

(b) The boiling point of water is 100°C. What is it in Fahrenheit?

(c) A comfortable room temperature is about 72°F. What is it in Celsius?

(d) The temperature in a domestic fridge should be about 36°F. What is this in Celsius?

(e) The melting point of iron is about 2800°F. What is this in Celsius?

(f) The temperature at the centre of the sun may be about 27 000 000°F. What is it in Celsius?

(g) In 1895 a temperature of ⁻17°F was recorded in Scotland. What is this in Celsius?

(h) In 1983 a temperature of ⁻128°F was recorded in Antarctica. What is this in Celsius?

(i) The freezing point of mercury is ⁻38.86°C. What is it in Fahrenheit?

(j) At what temperature are the measurements in degrees Celsius and degrees Fahrenheit exactly the same?

D2 Building bricks are supplied in cube-shaped packs on a pallet. The weight of a pack can be worked out using the formula:

$$w = 2000l^3$$

w is the weight of a pack in kilograms and l is the length of the side of the pack in metres.

Work out the weights of packs of bricks with side lengths

(a) 1.0 metres (b) 1.5 metres (c) 50 centimetres

D3 *ProPlanters* make large cubical lead planters. The weight of a planter, filled with earth, is given by the formula $w = 0.6S^2 + 2.2S^3$.

w is the weight of the full planter in tonnes and S is the length of the side of the planter in metres.

(a) Work out the weight of a full planter with a side of 1.1 metres.

(b) Work out the weight of a full planter that has a side of 1 m 50 cm.

(c) Four gardeners can lift about 300 kg between them. Could they lift a full 50 cm planter?

D4 The weight in kg that can be supported at the middle of an oak beam is given by the formula
$$w = \frac{60bd^2}{l}.$$
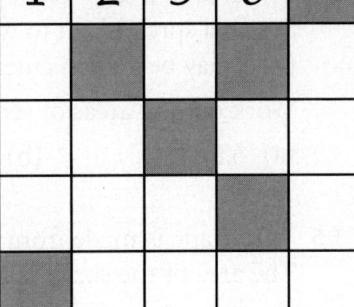

w stands for the weight in kg, and b, d and l for the breadth, depth and length of the beam in cm.

Calculate the load which can be supported by an oak beam 4 m long, 20 cm broad and 30 cm deep.

D5 A child drops a stone from the top of a cliff which is 80 m above the level of the sea below.

As it falls, the height of the stone above the sea can be calculated using the formula: $h = 80 - 5t^2$.

t is the time in seconds since the stone was dropped.
h is the height of the stone in metres above sea-level.

(a) Copy and complete this working to find the value of h when $t = 3$.

> When $t = 3$, $h = 80 - 5 \times 3^2$
> $= 80 - 5 \times 9 = 80 - 45 = $ ●

(b) Calculate the values of h when t is 0, 1, 2, and 4. Make a table of the results, like this.

t	0	1	2	3	4
h					

(c) Describe where the stone is when $t = 0$.

(d) Does the stone fall at the same speed all the time or does it speed up, or does it slow down?

(e) After how many seconds does the stone hit the sea?

D6 Copy this number grid puzzle onto squared paper. To solve the puzzle, first work out the value of each of the expressions below.

Then you have to fit the values you worked out into the grid puzzle.

To help you, the value of one of the expressions has been fitted into the grid for you.

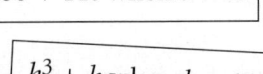

| $m^3 - 20m$ when $m = 14$ | $12a^2 + 3a$ when $a = 10$ |

| $\dfrac{r^2 + 1}{2}$ when $r = {}^-7$ | $100 + 11s$ when $s = {}^-5$ |

| $(2k)^3$ when $k = 4$ | $h^3 + h$ when $h = 11$ | $2(20 + j)$ when $j = {}^-3$ |

| $100q^2 + 4$ when $q = {}^-5$ | $3(6t - 1)$ when $t = 12$ | x^2 when $x = 6$ |

E *Using a spreadsheet*

A spreadsheet is ideal when you have a lot of calculations to do using a formula.
For example, this spreadsheet is set up to work out the areas of trapeziums.

	A	B	C	D	E
	Length of first parallel side (a)	Length of second parallel side (b)	Height (h)	Area	
1					
2					
3	8	12	4	=0.5*(A3+B3)*C3	
4					

Trapezium spreadsheet

The area of a trapezium is usually written as $\frac{1}{2}(a + b)h$.
Notice that we have to write the formula so that the spreadsheet understands it.

E1 Set up your spreadsheet to work out the areas of trapeziums where:

(a) $a = 6, b = 5, h = 10$ (b) $a = 7.6, b = 12.2, h = 9.8$

E2 This trapezium has an area of $100\,\text{cm}^2$.
Use your spreadsheet to find the
value of x to one decimal place.

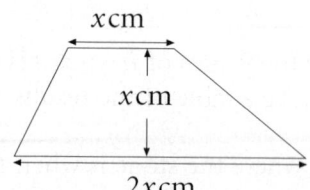

E3 Another trapezium has one parallel
side 3 cm longer than the other.
The height of the trapezium is the same
as the shorter side and its area is $50\,\text{cm}^2$.

Use a spreadsheet to work out the height of the trapezium to one decimal place.

E4 The area of a semi-circle is given by the expression $\frac{1}{2}\pi r^2$.
Set up a spreadsheet to work out the areas of semi-circles.
(You may be able to enter π as PI(); if not use 3.14159.)

Work out the areas of semi-circles with radii

(a) $6.5\,\text{cm}$ (b) $8.8\,\text{cm}$ (c) $0.2\,\text{km}$ (d) $100\,\text{m}$

E5 This shape is made from a square and a semi-circle.
The area of the shape is $200\,\text{cm}^2$.

Use a spreadsheet to work out the length of the side
of the square to one decimal place.

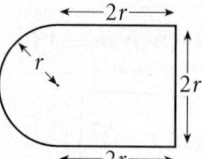

E6 This running track is made up of a square and two semi-circles.
The distance round the outside of the track must be 400 metres.

Use your spreadsheet to work out what the radius
of the semi-circles must be to the nearest centimetre.

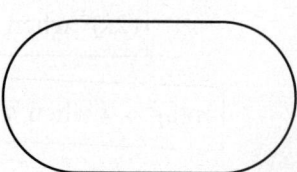

What you should have learned

◆ How to substitute into an expression involving cubes.

◆ How to substitute into more complex expressions, and those involving units.

◆ How to substitute numbers into expressions involving several letters.

Test yourself with these questions

T1 Calculate in your head the values of each of the following
when $a = 2$, $b = 4$, $c = {}^-1$.

(a) $a(b + c)$ (b) $ab - c$ (c) $\dfrac{ab}{c}$ (d) $\dfrac{b - 2c}{a}$ (e) ab^2

T2 Work out each of the following expressions without
a calculator when $p = \frac{1}{3}$, $q = \frac{1}{4}$ and $r = \frac{7}{8}$.

(a) $p + q$ (b) $3q$ (c) $3q - p$ (d) $p + q + r$ (e) $r - 2q$

T3 The area, A, of a trapezium can be written as $A = \frac{1}{2}(a + b)h$.
Work out the areas of each of these.

(a)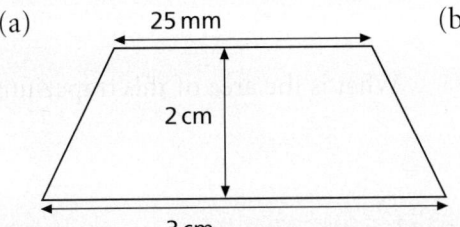

(b)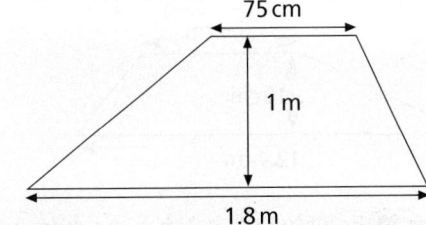

T4 The length of this metal rod is exactly 1 metre at 0°C.
When the temperature is T°C, its length, L metres,
is given by the formula $L = 1 + \dfrac{T}{30\,000}$.

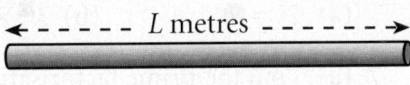

What is the length of the rod at

(a) 150°C (b) 600°C (c) ⁻90°C (d) ⁻200°C

T5 *Quattro* ponds are made up of a square and four semi-circles.
The radius of each semi-circle is r.

The area of a pond can be written as $2\pi r^2 + 4r^2$.
Work out the area of a *Quattro* pond when

(a) $r = 1$ (b) $r = 10$ (c) $r = 0.75$

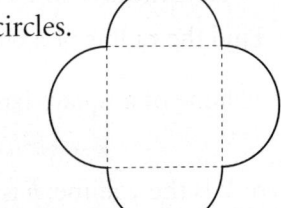

Review 2

Do not use a calculator for questions 1 to 7.

1 (a) What is the area of the whole parallelogram?

(b) Find the area of the triangle.

(c) (i) Write the area of the triangle as a
fraction of the area of the parallelogram.
Write it in its lowest terms.

(ii) Write this fraction as a decimal.

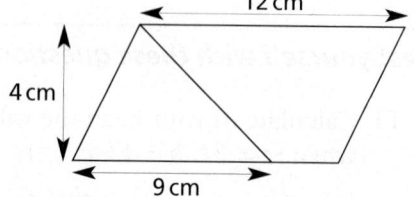

2 (a) What is the highest common factor of 36 and 27?

(b) (i) What is the lowest common multiple of 15 and 9?

(ii) Which is larger, $\frac{5}{9}$ or $\frac{7}{15}$?

(iii) Calculate $\frac{4}{9} - \frac{2}{15}$.

3 What is the value of $\frac{3n^2}{2}$ when $n = 4$?

4

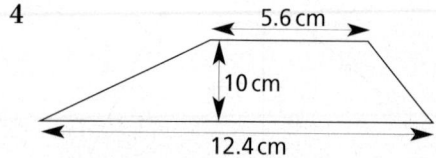

What is the area of this trapezium?

5 What is the value of $a + b + c$ when $a = \frac{1}{2}$, $b = \frac{2}{5}$ and $c = \frac{7}{10}$?
Write your answer as a mixed number in its lowest terms.

6 Write down the numbers missing from these statements.

(a) $2^5 = \blacksquare$ (b) $3^{\blacksquare} \times 2^3 = 72$ (c) $2^4 \times 2^5 = 2^{\blacksquare}$ (d) $7^6 \times 7^{\blacksquare} = 7^{10}$

7 (a) Find the prime factorisation of 300 and write it using index notation.

(b) Find the prime factorisation of 198 and write it using index notation.

(c) Find the highest common factor of 198 and 300.

8 (a) What is the area of a circle that has a radius of 5.2 cm, correct to one decimal place?

(b) Find the radius of a circle with area 100 cm², correct to one decimal place.

9 The volume of a square based pyramid is given by the formula
$$V = \frac{1}{3}b^2h$$
where V is the volume, b is the edge length of the square base and h is the height.
Find the volume of a square based pyramid where $b = 4.5$ cm and $h = 9.8$ cm.

Probability rules

This unit will help you to write probabilties as fractions and to list all the outcomes in different probability situations.

You will see that there are situations where it is only possible to get probability estimates from experiments and others where the probability can be calculated.

A *Relative frequency*

Going potty

If you hold up a cottage cheese or similar shaped carton at nose height and drop it, it can land in one of three ways:

Which way is it most likely to land?
Which way is the least likely?

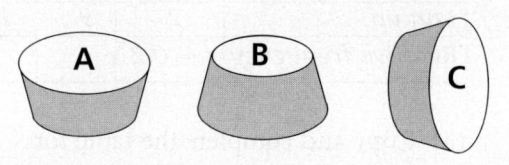

Trash

In a game show the winning contestant is shown three closed doors.
The host tells the contestant that behind one door is a new car, but behind the other two doors are trash cans.
The contestant is told to choose one door and they will win the prize behind it.

However, once the contestant has chosen, the host opens one of the **other** doors to reveal a trash can.

He then gives the contestant the choice of sticking with the original door they chose or changing to the other unopened door.

What should the contestant do?

Chinese dice

A set of Chinese dice consists of three dice numbered:-

      A: 6 6 2 2 2 2   *B: 5 5 5 5 1 1*   *C: 4 4 4 3 3 3*

One player chooses a dice to play with.

Then the second player chooses a dice to play with.

Both players roll their dice and the highest score wins.

It is said that by choosing their dice second a player can have a higher chance of winning. Is this true?

The **relative frequency** of an event happening in a probability experiment is:

$$\frac{\text{the number of times the event occurs}}{\text{the total number of trials}}$$

It gives an estimate of the probability of the event happening.

If a drawing pin was dropped 200 times and it landed point up 130 times, then

The relative frequency of the pin landing point up $= \frac{130}{200} = \frac{13}{20} = 0.65$

A1 Martin carries out an experiment dropping pieces of toast to see if they land 'jam-up' or 'jam-down'. Here are the results of his experiment.

Total trials	10	20	40	50	80	100
Jam up	2	9	17	19	30	38
Relative frequency	$\frac{2}{10}$ = 0.2					

(a) Copy and complete the table for Martin's experiment.

(b) Copy and complete this graph for the relative frequencies in Martin's experiment.

(c) From the results of Martin's experiment would you say that a piece of toast was more likely to land 'jam-up' or 'jam-down'?

(d) How good would you say Martin's estimate of the probability of a piece of toast landing 'jam-up' is after 100 trials?

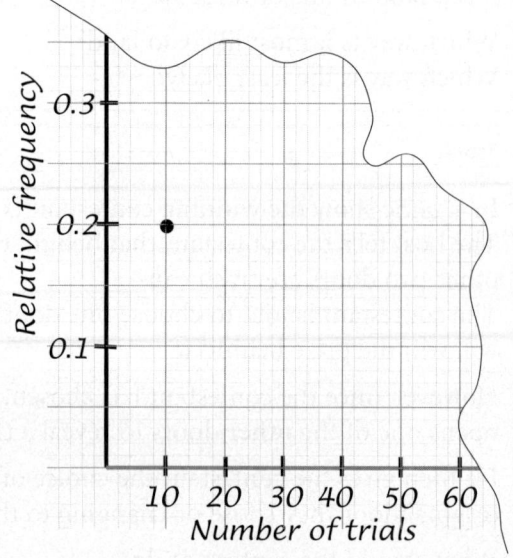

A2 Kirsten is recording which direction cars go when they reach a fork in the road near her school. She writes right (R) or left (L) for each car.
Here are her results:

R R L R L R R L L L   R R R L L R L R L R   L R L L R R R R R L

L R L R L R R R L R   R L R R R L L R R R

(a) What is the relative frequency of cars turning right from Kirsten's results, after

  (i) 20 results   (ii) 40 results   (iii) 50 results

(b) If 800 cars a week drive across this fork, roughly how many would you expect to go right?

(c) How reliable do you think Kirsten's estimate of the probability of going right is?

B *Equally likely outcomes*

In working out probabilities it is often necessary to assume that different possibilities or **outcomes** are all just as likely.

The probability of a particular outcome or one of a set of outcomes occurring is found by:

$$\text{probability of this outcome} = \frac{\text{number of outcomes in this set}}{\text{total number of equally likely outcomes}}$$

Also: probability that one of this set does not occur = 1 − probability one of set occurs.

If there are 3 red beads and 5 black beads in a bag, and each bead is equally likely to be chosen.

Probability of choosing red = $\frac{3}{8}$

Probability of not choosing red = $1 - \frac{3}{8} = \frac{5}{8}$

If we were looking at the probability of a family of four children all being boys we might assume that each child is equally likely to be a boy or girl. In reality the probability of a child being born a boy is $\frac{52}{100}$.

B1 Which of these sets of outcomes are equally likely?

(a) Having a birthday in a particular month, (Jan, Feb, March ....)

(b) A particular Lottery ball coming up first (1, 2, 3, 4, ...49)

(c) A card from a pack being of a particular suit (clubs, diamonds, hearts, spades)

(d) The next car to pass your window being of a particular colour (red, black, blue.....)

B2 A pack of 24 cards has the numbers 1 to 24 marked on them.
The pack is shuffled and one card is chosen from the pack.
What is the probability the number on the card is:

(a) even   (b) a square number   (c) a triangle number

(d) 5 or less   (e) a prime number   (f) not a prime number

B3 A box contains 3 red buttons, 4 blue buttons and 5 gold buttons.
A button is chosen at random and each button is equally likely to be chosen.
Find the probability that a button chosen is:

(a) red   (b) gold   (c) not blue

B4 This table shows information about a group of children.

	Boys	Girls
Blue eyes	3	6
Brown eyes	12	9

(a) How many children are in the group altogether?

(b) If a child is chosen from the group at random what is the probability of that child

(i) having blue eyes   (ii) being a boy?

(c) If a boy is chosen from the group what is the probability that he has blue eyes?

C *Listing outcomes*

If two or more trials are combined in probability situations, a list of all the possible outcomes can be helpful. The outcomes in each trial must be equally likely.

Worked example

If a coin is flipped and an ordinary dice rolled, what is the probability of getting a head and a number greater than 4?

The outcomes can be listed: H1   H2   H3   H4   (H5)   (H6)

T1   T2   T3   T4   T5   T6

There are 12 outcomes altogether.
Two outcomes are a head and a number greater than 4 (ringed).

So the probability of flipping a head and rolling a number greater than $4 = \frac{2}{12} = \frac{1}{6}$.

C1 If a coin is flipped and an ordinary dice is rolled, what is the probability of getting

(a) a tail and a number less than 5

(b) a head and an even number

(c) a head and an even number or a tail and an odd number

(d) a tail and a number other than 1?

C2 (a) List all the outcomes when a 5p, 10p and £1 coin are flipped. How many outcomes are there altogether?

(b) What is the probability of all 3 coins showing a head?

(c) What is the probability of all 3 coins showing the same face?

(d) What is the probability of 2 or more coins showing a tail?

(e) What is the probability that there are more heads than tails showing?

C3 A fortune teller advises customers about their romances using two small packs of cards.

Pack A: Contains the Queens of Hearts, Diamonds, Spades and Clubs from a pack.
Pack B: Contains the Kings of Hearts, Diamonds, Spades and Clubs from a pack and the Jack of Hearts.

The fortune teller turns over one card from each pack and what she tells people depends on the two cards showing:

- If they show the Queen and King of Hearts it is 'a match made in heaven'

- If they show the King and Queen of the same suit 'this will be long lasting'

- If the King and Queen are of different suits then 'this will not last long'

- If it is a red King and black Queen or vice versa it will be a 'stormy end'

- If the Jack of Hearts turns up then 'someone new will appear very soon'

(a) List all the outcomes for the two cards showing.

(b) Find the probability of the fortune teller saying each of the statements above.

C4 Tim writes the 3 letters of his name on pieces of card.
He turns the cards over and shuffles them.
He then turns them face up in a row.

What is the probability the cards spell his name?

C5 Credit cards often have a PIN (Personal Identification
Number) for security.
These usually consist of 4 digit numbers

(a) Sarah knows that her PIN uses the numbers
1, 9, 6, 7 but cannot remember the order.
Copy and complete this list of the possible
PIN numbers that she might have.

PLEASE ENTER YOUR PIN

♦ ♦ ♦ ♦

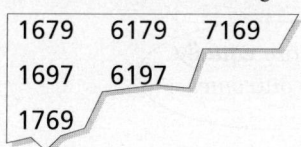

1679	6179	7169
1697	6197	
1769		

(b) If Sarah tries to guess her number by using the numbers 1,9,6,7 at random,
what is the probability she guesses right first time?

(c) Elizabeth cannot remember her number either, but she knows it contains the
digits 8, 3, 5, 5.
What is the probability Elizabeth guesses her number correctly by putting in the
digits 8, 3, 5, 5 in a random order?

Honeycomb

This bee wants to go from the top cell to the bottom
row going down the honeycomb cell by cell.
From each cell he can go either to the Left one
below or the Right one below.
He is just as likely to go Left or Right each time.

This route shows him going Right, Left, Right to
arrive at cell 2.

List all the possible routes he can go to the bottom.

Find the probability of finishing in each of the cells?

Investigate for different honeycombs.

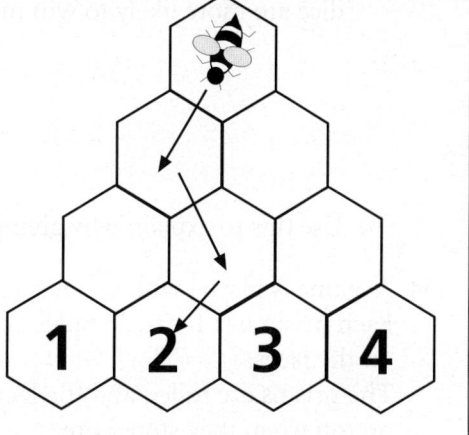

D Using grids

Grids can be a useful way of showing all the outcomes in probability situations.

In the Chinese Dice problem all the outcomes can be shown by listing all the possibilities in a grid.

For dice A playing dice B the grid would look like:

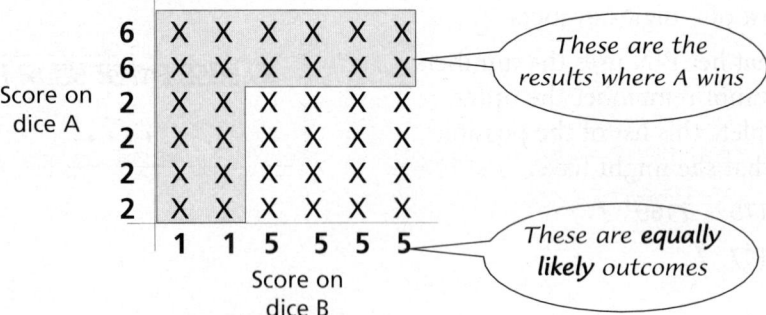

This shows that the probability of A beating B is $\frac{20}{36} = \frac{5}{9}$

The probability of B winning $= 1 - \frac{5}{9} = \frac{4}{9}$

D1 (a) Draw a grid to show all the outcomes when dice B is played against dice C in Chinese Dice.

(b) Shade the area on the grid which shows where dice B beats dice C.

(c) What is the probability of dice B beating dice C?

(d) What is the probability of dice C beating dice B?

D2 (a) Draw a grid to show the outcomes of dice A playing dice C.

(b) Use your grid to find the probabilities of A and C winning.

D3 (a) Copy and complete this diagram to show which dice are more likely to win in Chinese dice.

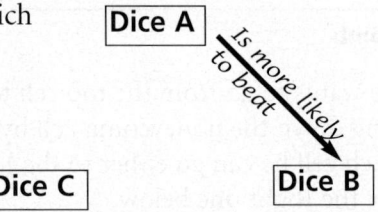

(b) Use this to explain why giving the other player first choice of dice helps you to win.

D4 A game is played with two hexagonal prisms.
Each prism has 1 star, 2 apples and 3 bananas on the faces.
The prisms are rolled and the faces showing on top when they stop score.

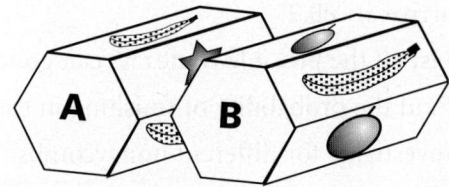

(a) Use a grid to show all the outcomes of rolling the two prisms.

(b) Use your grid to help you calculate the probability of

　(i)　both top faces showing a star

　(ii)　only one of the top faces showing a star

　(iii)　both top faces showing the same symbol

　(iv)　no face showing a star

D5 A game is played with two spinners numbered 1 to 5.

The player spinning adds together the two scores that the spinner lands on.

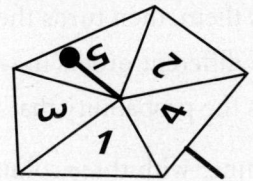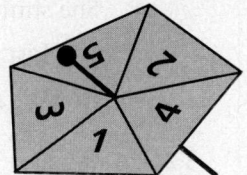

(a) Copy and complete this grid showing the **total** scores that can be obtained using the two spinners.

Total scores

5					
4					
3	4	5			
2	3	4	5		
1	2	3	4	5	6
	1	**2**	**3**	**4**	**5**

Score on white spinner

Score on red spinner

(b) Use your grid to find the probability of a player

　(i)　scoring a total of 10

　(ii)　scoring a total of 5 or more

　(iii)　scoring a total less than 5

　(iv)　scoring a total which is a prime number

　(v)　getting the same score on both spinners

　(vi)　getting a higher score on the red spinner than the white one

D6 In another game using the same two spinners as in C5, players score the **difference** between the scores on the two spinners.

(a) Draw a grid showing the differences that can be obtained from the spinners.

(b) Use your grid to find the probability of a player

　(i)　scoring zero　　　　　(ii)　scoring more than 2

　(iii)　scoring less than 2　　(iv)　scoring 5

Test yourself with these questions

T1 (a) Graham has 9 cards numbered 1, 2, 3, 4, 5, 6, 7, 8, 9.
He picks one at random. What is the probability that he picks

    (i) 5   (ii) 10   (iii) a multiple of 4?

  (b) Kira has 3 cards, numbered 1, 2, 3.
She shuffles them, then turns them over one by one.

    (i) List the different orders in which the three cards can appear.

    (ii) What is the probability that the three cards do **not** appear in the order 1, 2, 3?

T2 Lyra made a spinner with three colours, yellow, blue and red.
She tested it by spinning it 500 times.
Her results were: 234 landed on yellow   167 landed on blue   99 landed on red.

  (a) Estimate the probability of the spinner landing on yellow.

  (b) She then spun the spinner 100 times.
How many times would you expect the spinner to land on yellow?

T3 These are two sets of cards.

SET P | 3 | | 4 | | 5 | **SET Q** | 0 | | 2 | | 4 |

A card is taken at random from set P and another from set Q.

  (a) List all the possible outcomes

The numbers on the two cards taken are added together.

  (b) What is the probability of getting a total that is an odd number?

  (c) What is the probability of getting a total that is an even number?

T4 These two fair spinners are used for a game.
The score is the **difference** of the numbers
the spinners land on.

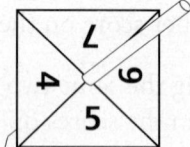
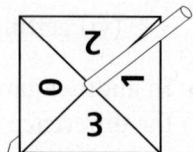

  (a) Copy and complete this table to show all
the possible scores for the two spinners

  (b) What is the probability of the score
being a prime number?

	4	5	6	7
0				
1	3			
2			4	
3				

12 Changing the subject 1

A Planting

Stella is a garden designer who plants ornamental vegetable beds.
She plants rows of artichokes.
She plants marigolds by her artichokes to keep away pests.

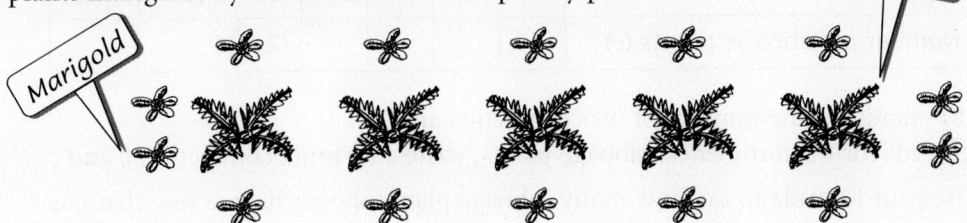

Stella plants marigolds around each row of artichokes in the pattern shown above.
This table shows the number of marigolds for rows of artichokes of various lengths.

Number of artichokes (a)	1	2	3	4	5
Number of marigolds (m)	6	8	10	12	14

A rule for finding the number of marigolds needed is

   number of marigolds = number of artichokes × 2 + 4

If a stands for the number of artichokes and m stands for the number of marigolds
we can write this rule as a **formula** using letters:

   $m = a \times 2 + 4$ or even shorter as $m = 2a + 4$.

A1 Use the formula to complete these shopping lists. Each is for one row of artichokes.

(a)
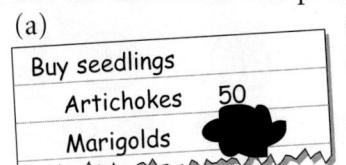
Buy seedlings
Artichokes   50
Marigolds

(b)
Plant list
Artichokes   75
Marigolds

(c)
Order from Joad's
Artichoke plants   15
Marigold plants

A2 This is another order for one row of artichokes.
We know the number of marigolds it needs,
but we need to find the number of artichokes.

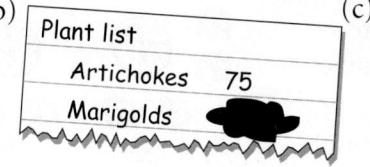
Order from Brown's
Artichoke plants
Marigold plants   80

Put m = 80 in the formula m = 2a + 4.
Solve the equation you get to find a.

How many artichoke plants are needed?

A3

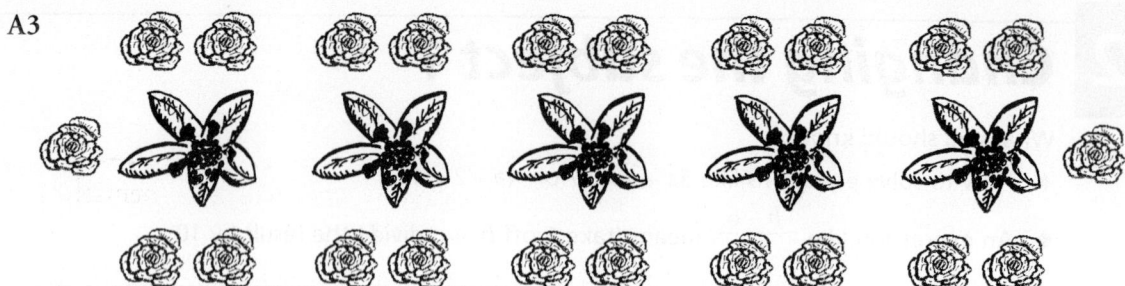

Stella also plants rows of broccoli.
She puts ornamental cabbages above and below each row of broccoli
and at the ends in the pattern shown above.

(a) Copy and complete this table for her broccoli and cabbage rows.

Number of broccoli plants (*b*)	1	2	3	4	5			10	100
Number of cabbage plants (*c*)					22				

(b) If *b* stands for the number of broccoli plants and
c stands for the number of cabbage plants, write a formula connecting *b* and *c*.

(c) Use your formula to say how many cabbage plants she needs in a row that has
(i) 15 broccoli plants   (ii) 25 broccoli plants

This is an order for one row of broccoli plants.
We need to find the number of
broccoli plants to order.

(d) Put *c* = 90 in your formula.
Solve the equation you get to find *b*.

(e) How many broccoli plants does Stella need?

> Order from Brown's
> Broccoli plants
> Cabbage plants   90

A4 Stella grows Spanish onions and
garlic in patterns like this.

Suppose *s* stands for the number
of Spanish onions in a row
and *g* stands for the number
of garlic bulbs.

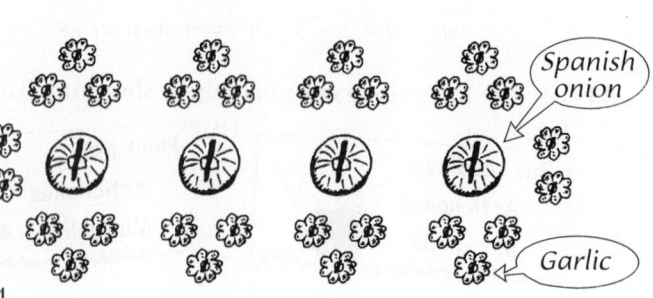

Spanish onion

Garlic

(a) Explain why the formula
connecting *s* and *g* is   *g* = 6*s* + 4.

(b) How many garlic bulbs does Stella need if she plants
50 Spanish onions in one row?

(c) For one row of onions, Stella plants 100 garlic bulbs.
How many Spanish onions are in the row?

B New formulas from old

For Stella's onion and garlic rows, the formula $g = 6s + 4$
connects the letters g and s.

We can find s when $g = 100$
like this:

$$100 = 6s + 4$$
$$96 = 6s$$
$$16 = s$$

take 4 from both sides

divide both sides by 6

We can work with the letters themselves
instead of numbers.

$$g = 6s + 4$$
$$g - 4 = 6s$$
$$\frac{g-4}{6} = s, \text{ or } s = \frac{g-4}{6}$$

Each time, we do the same thing to both sides of the formula.
We call this 're-arranging the formula to make s the subject' or
'making s the subject of the formula'.

Now we can find the value of s easily for different values of g.

We can check the re-arrangement by using values of g and s.
For example, when $s = 10$ in the original formula, $g = 6 \times 10 + 4 = 64$.
Now check this in the re-arranged formula.

When $g = 64$, $s = \frac{64-4}{6} = \frac{60}{6} = 10$ which checks.

B1 For Stella's onion and garlic pattern, a formula is $s = \frac{g-4}{6}$
s stands for the number of Spanish onions in a row
and g stands for the number of garlic bulbs.

(a) If $g = 160$, what is s?

(b) When $g = 400$, what is s?

(c) There are 82 garlic bulbs in one row.
How many onions are there in this row?

B2 Stella plants red and white onions in rows like this.

Suppose r stands for the number of red onions
and w stands for the number of white onions.

(a) Explain why $w = 3r + 2$ in this pattern.

(b) What is the value of w when $r = 20$?

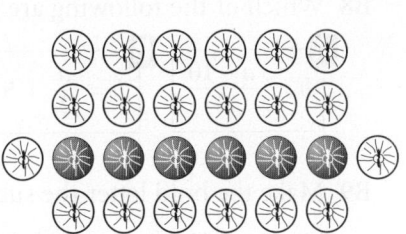

(c) Copy and complete this working
to make r the subject of the formula.

(d) What is the value of r when $w = 50$?

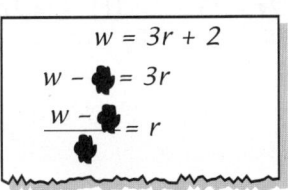

$$w = 3r + 2$$
$$w - \clubsuit = 3r$$
$$\frac{w - \clubsuit}{\clubsuit} = r$$

(e) How many red onions are there in a row
that has 68 white onions in it?
Check your answer works in the formula for w.

B3 For this pattern of red and white onions $w = 2r + 4$.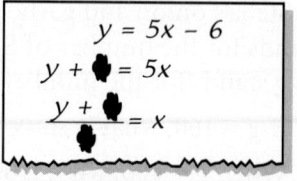

    (a) Rearrange the formula $w = 2r + 4$
        to make r the subject.

    (b) Work out r when $w = 88$.

    (c) How many red onions are there in a row with 128 white?

B4 (a) In the formula $f = 3g + 2$, find f when $g = 10$.

    (b) Make g the subject of the formula $f = 3g + 2$.

    (c) Check your re-arrangement is correct by substituting
        the value of f from part (a) into your new formula.

B5 (a) Rearrange the formula $s = 5t + 1$ to make t the subject.

    (b) Find a pair of values of s and t that fit the original formula, $s = 5t + 1$.
        Use this pair of values to check that your re-arrangement is correct.

B6 Make the bold letter the subject of each of these formulas.
For each one, check your re-arrangement by using a pair of values
that fit the original formula.

    (a) $b = 8\mathbf{w} + 7$ (b) $u = 5\mathbf{v} + 2$ (c) $g = 6\mathbf{d}$ (d) $y = 12 + 3\mathbf{x}$

    (e) $t = 3\mathbf{b} + 5$ (f) $f = 8 + 3\mathbf{d}$ (g) $h = \mathbf{k} + 5$ (h) $w = 7\mathbf{d} + 1$

B7 (a) Copy and complete this working to make x
        the subject of the formula $y = 5x - 6$

    (b) Use your new formula to find x when $y = 129$.

    (c) Substitute $y = 129$ and the value of x you found
        in part (b) in the original formula to check
        that your re-arrangement is correct.

$$y = 5x - 6$$
$$y + \bullet = 5x$$
$$\frac{y + \bullet}{\bullet} = x$$

B8 Which of the following are correct re-arrangements of $a = 2b - 10$?

A $b = \dfrac{a - 10}{2}$ **B** $b = \dfrac{a}{2} + 5$ **C** $b = \dfrac{a + 10}{2}$ **D** $b = \dfrac{a + 2}{10}$ **E** $b = \dfrac{a - 2}{10}$ **F** $b = \dfrac{10 + a}{2}$

B9 Make the bold letter the subject of each of these formulas.

    (a) $a = 8\mathbf{w} - 6$ (b) $b = 4\mathbf{h} - 1$ (c) $h = 2\mathbf{f} - 2$ (d) $y = \mathbf{x} - 15$

    (e) $z = 2\mathbf{r} - 15$ (f) $k = 2\mathbf{d} - 3$ (g) $b = \mathbf{g} - 5$ (h) $l = 2\mathbf{m} - 1$

B10 Here are 8 formulas.
Find four matching pairs
of equivalent formulas.

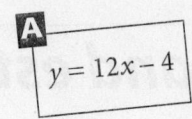 **A** $y = 12x - 4$

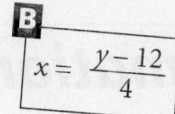 **B** $x = \dfrac{y - 12}{4}$

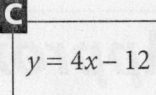 **C** $y = 4x - 12$

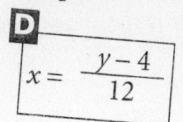 **D** $x = \dfrac{y - 4}{12}$

E $y = 4x + 12$

F $x = \dfrac{y + 4}{12}$

G $y = 12x + 4$

H $x = \dfrac{y + 12}{4}$

B11 Re-arrange each of these formulas to make the bold letter the subject.

(a) $a = 30 + 3\boldsymbol{b}$ (b) $s = 2\boldsymbol{t} - 40$ (c) $t = 12\boldsymbol{g} - 60$ (d) $f = 3\boldsymbol{b} + 12$

(e) $y = 12 + 8\boldsymbol{x}$ (f) $r = 5\boldsymbol{s} - 20$ (g) $a = 3\boldsymbol{b}$ (h) $v = 7\boldsymbol{u} - 10$

(i) $y = 35 + \boldsymbol{x}$ (j) $8 + 4\boldsymbol{j} = d$ (k) $k = 8\boldsymbol{j} - 45$ (l) $7\boldsymbol{z} - 1 = w$

What you should have learned

How to change the subject of formulas with letters and numbers in them, for example

◆ making d the subject of $a = 4d + 5$ ◆ making k the subject of $h = 2k - 3$

Test yourself with these questions

T1 Make the bold letter the subject of each of these formulas.

(a) $a = 6\boldsymbol{r} + 8$ (b) $b = 4\boldsymbol{s} + 6$ (c) $c = 12 + 5\boldsymbol{t}$ (d) $d = 8 + 4\boldsymbol{u}$

T2 (a) Copy and complete this working to make m
the subject of the formula $n = 3m - 2$.

(b) Find a pair of values of n and m
that fit the formula $n = 3m - 2$.

(c) Use this pair of values to check
your rearrangement.

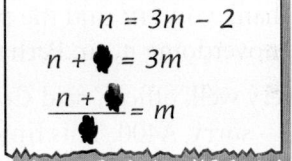

$n = 3m - 2$

$n + \bullet = 3m$

$\dfrac{n + \bullet}{\bullet} = m$

T3 Which of these rearrangements of $y = 2x - 3$ are correct?

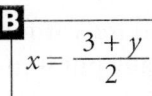 **A** $x = \dfrac{3 - y}{2}$ **B** $x = \dfrac{3 + y}{2}$ **C** $x = \dfrac{y - 3}{2}$

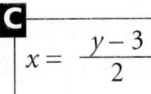 **D** $x = \dfrac{y + 3}{2}$ **E** $x = \dfrac{y - 2}{3}$ **F** $x = \dfrac{y + 2}{3}$

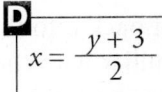 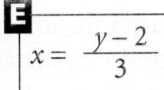

T4 Rearrange each of these formulas to make the bold letter the subject.

(a) $e = 4\boldsymbol{v} - 12$ (b) $f = 2\boldsymbol{w} - 15$ (c) $g = \boldsymbol{x} - 12$ (d) $h = 5\boldsymbol{y} - 10$

T5 Make the bold letter the subject of each formula.

(a) $n = 8 + 5\boldsymbol{t}$ (b) $s = 4\boldsymbol{t} - 7$ (c) $y = 2\boldsymbol{x} + 12$ (d) $m = 3\boldsymbol{u} - 5$

(e) $j = 5\boldsymbol{v} + 12$ (f) $f = 15 + 5\boldsymbol{w}$ (g) $k = 3\boldsymbol{x}$ (h) $8\boldsymbol{y} + 4 = b$

13 Approximation and estimation

This work will help you

◆ round numbers to the nearest hundred, thousand, and so on
◆ round to a given number of decimal places
◆ round to a given number of significant figures
◆ estimate the result of a calculation by rounding

A Rounding numbers

'This has to be the worst day of my life,' said Colin as he emerged from the wreckage of his car.

'I can understand you feeling like that,' said the police officer. 'Could you please give me as accurate an account as you can of the events leading up to the accident.'

'Well,' said Colin, 'I woke up at 32 minutes 43.8 seconds past 6. As always, I weighed myself and found that I was 72.451 kg, that's 0.078 kg more than yesterday. I put on my shirt, collar size 31.3264 cm, and trousers, waist size 71.5093 cm. I went down and had breakfast. I had 74.832 g of cornflakes with 0.32762 litres of milk. I left the house at 7:14:26.3 and got into my car.'

'What model was the car, sir?' asked the officer. 'It's hard to tell from what's left of it.'

'It's a Ferraghini 3.4782 litre, capable of a top speed of 163.629 m.p.h.'

'Thank you, sir,' said the policeman. 'Now I did ask you to be accurate, but you are rather overdoing it, sir. Perhaps you could round off the numbers from now on.'

'Very well, officer,' said Colin. 'I'll do my best. Anyway, I usually drive to work on the A382 – sorry, A400. This time as the weather was fine I went by the country route, the B2864 total that – sorry, B3000. I passed a number 148 bus – sorry, 150 bus – and suddenly realised that I had left my case at my friend's house yesterday. He lives at 318 – sorry, 300 – Elm Road, but I didn't have time to go there. So I stopped at a phone box and rang him up. His number is 347 2846 – sorry, 350 0000.

'He wasn't in. So I got back into the car. I've only had the car for 4 days – sorry, 0 days – and I wasn't familiar with the gears. I accidentally put it in reverse and collided at full speed with the 150 bus as it came round the corner.'

A1 Round 4386 to the nearest (a) thousand (b) hundred (c) ten

A2 Round 2396 to the nearest (a) thousand (b) hundred (c) ten

A3 Round 40 789 to the nearest (a) ten (b) hundred (c) thousand

A4 Round (a) 32 096 to the nearest ten (b) 48 607 to the nearest hundred

B Large numbers

This table shows the population of Greater London at each census in the early part of the 20th century.

Greater London

Year	Population
1901	6 586 269
1911	7 225 946
1921	7 488 382
1931	8 215 673

B1 Round the 1901 population to the nearest
  (a) hundred thousand   (b) ten thousand
  (c) thousand   (d) hundred

B2 Round the 1911 population to the nearest
  (a) hundred thousand   (b) thousand

B3 Round the 1921 population to the nearest
  (a) ten thousand   (b) hundred

B4 Round the 1931 population to the nearest
  (a) hundred thousand   (b) thousand

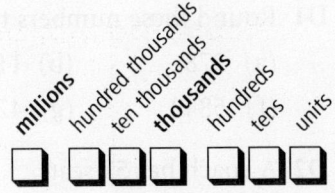

This diagram may help you.

B5 Round each of the 1901, 1911, and 1921 populations to the nearest million.
    Why is it not a good idea to round them like this?

C Rounding decimals

Worked example	
Round 4.2763 to 2 decimal places	4.2763 is between 4.27 and 4.28. Round up if the digit in the next decimal place is 5 or more. Here it is 6, so round up to **4.28**

C1 Round each of these numbers to one decimal place (1 d.p.).
  (a) 48.32   (b) 8.754   (c) 0.4503   (d) 23.962   (e) 70.0413

C2 Round each of these numbers to 2 d.p.
  (a) 3.9563   (b) 0.08732   (c) 0.1659   (d) 3.5031   (e) 143.6395

C3 Round (a) 3.4783 to 1 d.p.   (b) 4.08312 to 2 d.p.   (c) 8.05723 to 3 d.p.
          (d) 0.79621 to 2 d.p.   (e) 0.067843 to 3 d.p.   (f) 10.8956 to 1 d.p.

C4 Do these on a calculator and round each answer to two decimal places.
  (a) $2.65 \div 3.47$ (b) $4.818 \div 0.357$ (c) 0.159×0.357 (d) 16.77×0.167
  (e) 4.87×0.913 (f) $3.007 \div 27.55$ (g) $0.2619 \div 0.125$ (h) $1.169 \div 0.3894$

C5 (a) Calculate $2.467 \div 6.123$, giving the answer to 3 d.p.
    (b) Calculate 3.348×4.17, giving the answer to 1 d.p.
    (c) Calculate 0.5913^2, giving the answer to 3 d.p.

D *Rounding to one significant figure: whole numbers*

The first significant figure in a number is
the figure with the highest value.

| 34 168 | 286 | 5 876 672 |

We can round to **one significant figure:** 30 000   300   6 000 000

D1 Round these numbers to one significant figure.

  (a) 278 (b) 11 328 (c) 5418 (d) 863 (e) 304 657

  (f) 5842 (g) 421 987 (h) 27 083 (i) 800 264 (j) 961

D2 A coach has 57 seats.
   This is how Jack estimates the number of seats
   in 32 coaches.

   Complete his estimate.

> Round the numbers to one significant figure:
>   57 becomes 60   32 becomes 30
>
> So 57 × 32 is roughly

D3 Work out a rough estimate for each of these.

  (a) 78 × 21 (b) 42 × 39 (c) 63 × 22 (d) 48 × 19 (e) 27 × 44

  (f) 291 × 33 (g) 58 × 188 (h) 37 × 487 (i) 81 × 77 (j) 196 × 207

D4 Tickets for a concert cost £29.50 each.
   Estimate the total amount taken if 403 tickets are sold.

E *Rounding to one significant figure: decimals*

The first significant figure is the first non-zero figure you come to
working along from left to right.

| 2.34 6 | 0.004 83 | 0.000 029 3 |

Rounding to one significant figure, we get: 2   0.005   0.000 03

E1 Round these numbers to one significant figure.

  (a) 7.537 (b) 0.8851 (c) 0.04287 (d) 0.06763 (e) 0.003196

  (f) 0.08853 (g) 0.000475 (h) 26.46 (i) 0.00364 (j) 4.0075

E2 Round these numbers to one significant figure.

  (a) 17.507 (b) 0.0334 (c) 0.004636 (d) 0.01005 (e) 0.008089

  (f) 347.07 (g) 0.68846 (h) 2.0775 (i) 0.000767 (j) 0.9858

Worked example

Estimate the answer to 0.364×516.

0.364×516

Round to one significant figure: $0.4 \quad \times 500 = \mathbf{200}$

$4 \times 5 = 20$
$4 \times 50 = 200$
$4 \times 500 = 2000$
$0.4 \times 500 = 200$

E3 Estimate the answer to each of these.
(a) 7.2×0.23 (b) 0.48×3.13 (c) 0.27×0.41 (d) 0.186×176 (e) 68.2×0.27
(f) 2.84×0.32 (g) 378×1.77 (h) 0.471×42.7 (i) 8.71×0.031 (j) 53.2×0.97

E4 Estimate the cost of
(a) 0.475 kg of Stilton
(b) 0.856 kg of Brie
(c) 32.5 kg of Cheddar

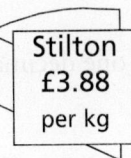
Stilton
£3.88
per kg

Brie
£5.15
per kg

Cheddar
£1.96
per kg

F *Rounding to two or more significant figures*

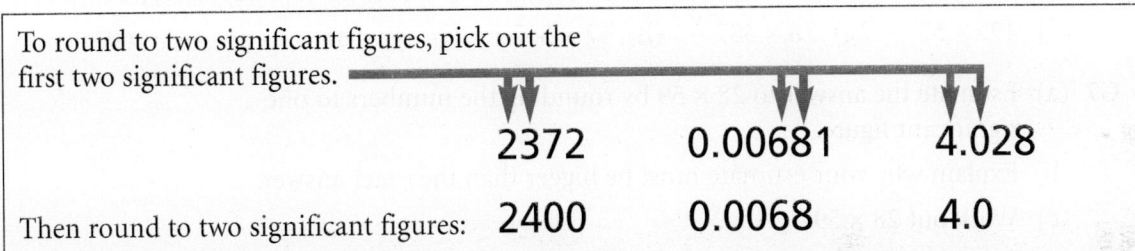

To round to two significant figures, pick out the
first two significant figures.

	2372	0.00681	4.028
Then round to two significant figures:	2400	0.0068	4.0

F1 Round these numbers to two significant figures.
(a) 4628 (b) 13 752 (c) 20 984 (d) 378 (e) 341 543
(f) 29 741 (g) 5381 (h) 638 102 (i) 5 328 613 (j) 9642

F2 Round these to three significant figures
(a) 7126 (b) 43 209 (c) 60 573 (d) 3328 (e) 28 031
(f) 159 762 (g) 3612 (h) 600 813 (i) 4 136 442 (j) 10 527

F3 Round these numbers to two significant figures.
(a) 0.053 14 (b) 2.387 (c) 3.075 (d) 0.004 284 (e) 0.6696
(f) 0.000 485 1 (g) 51.852 (h) 0.067 52 (i) 6.0238 (j) 0.000 974

F4 Round these numbers to three significant figures.
(a) 0.023 74 (b) 8.189 (c) 41.023 (d) 0.006 414 (e) 0.7791
(f) 0.000 155 8 (g) 254.756 (h) 0.080 284 (i) 12.1238 (j) 0.006 918

TG

G Mixed questions

Do not use a calculator for the questions on this page.

G1 Round 346 718 to

(a) the nearest thousand (b) the nearest hundred (c) one significant figure

G2 Round 2 886 032 to

(a) the nearest million (b) the nearest ten thousand (c) one significant figure

G3 Round 7356.087 to

(a) the nearest hundred (b) one decimal place (c) one significant figure

G4 Round 2.4692 to

(a) the nearest whole number (b) one decimal place (c) two decimal places

G5 Round 0.058 21 to

(a) two decimal places (b) two significant figures (c) three significant figures

G6 Estimate the answer to each of these by rounding the numbers to one significant figure.

(a) 38×51 (b) 62×29 (c) 83×42 (d) 78×59 (e) 67×84

(f) 495×53 (g) 78×387 (h) 47×567 (i) 91×86 (j) 493×603

G7 (a) Estimate the answer to 28×59 by rounding the numbers to one significant figure.

(b) Explain why your estimate must be bigger than the exact answer.

(c) Work out 28×59.

G8 Pam works 28 hours a week. She is paid £6.85 per hour.

(a) Estimate how much she earns in a week.

(b) Estimate how much she earns in 31 weeks.

G9 Steve worked for 48 hours and was paid £290.40 altogether.
Estimate how much he was paid for each hour.

G10 Estimate the answer to each of these.

(a) 5.2×0.83 (b) 0.38×7.23 (c) 0.57×0.82 (d) 0.384×478 (e) 98.2×0.37

(f) 4.74×0.12 (g) 528×2.97 (h) 0.361×23.7 (i) 7.81×0.061 (j) 43.2×0.851

G11 Do these on a calculator and round each answer as shown.

(a) 4.375×0.853 Round the answer to 2 decimal places.

(b) $12.9 \div 0.0624$ Round the answer to 3 decimal places.

(c) $0.871 \div 4.271$ Round the answer to 3 decimal places.

G12 Round each answer to 3 significant figures.

(a) $4.65 \div 3.27$ (b) $6.873 \div 0.552$ (c) 0.349×0.254 (d) 36.27×0.368

(e) 2.47×0.814 (f) $7.052 \div 37.45$ (g) $0.2832 \div 0.168$ (h) $1.069 \div 0.3291$

G13 Round each answer to 3 significant figures.

(a) $(2.845 + 0.487) \times 0.864$

(b) $\dfrac{11.875 - 2.608}{0.427}$

(c) $0.425 + \dfrac{2.653}{5.175}$

(d) $\dfrac{5.375 + 2.634}{9.304 - 1.668}$

G14 A rectangular room measures 31.2 m by 42.3 m.

(a) Estimate the area of the room by rounding the measurements to one significant figure.

(b) Is your estimate bigger or smaller than the actual area? Explain how you can tell.

(b) Calculate the actual area, giving your answer to 3 significant figures.

Test yourself with these questions

T1 Round 2085.19 to

(a) the nearest hundred

(b) one decimal place

(c) one significant figure

(d) three significant figures

T2 Estimate the answer to each of these.

(a) 27.8×0.93 (b) 0.084×61.1 (c) 0.47×0.31 (d) 8.387×218 (e) 0.962×0.48

T3 In this question you must NOT use a calculator.
You must show all your working.

Tom buys 67 cameras at £312 each.

(a) Work out the total cost.

(b) Write down two numbers you could use to get an approximate answer to your calculation.   [Edexcel]

T4 Sally estimates the value of $\dfrac{42.8 \times 63.7}{285}$ to be 8.

Write down three numbers Sally could use to get her estimate: $\dfrac{\ldots\ldots \times \ldots\ldots}{\ldots\ldots}$

[Edexcel]

T5 (a) Write down 45.3476 correct to 3 significant figures.

(b) Write down 7462 correct to 2 significant figures.   [WJEC]

14 Brackets and equations

You should know how to

◆ Simplify an expression such as $1 - 2x + 6 - 3x$ to give $7 - 5x$

◆ Multiply out brackets such as $2(3x + 4)$ to give $6x + 8$

◆ Simplify divisions such as $\dfrac{8n - 2}{2}$ to give $4n - 1$

◆ Solve linear equations such as $6 + 3x = 10 - x$

You will learn how to

◆ Add and subtract expressions in brackets such as $(5 + 4x) - (3x + 4)$

◆ Simplify more complex expressions such as $4(3x - 5) - 2(4x - 1)$

◆ Use algebra to prove statements like 'The result for this puzzle will always be 4'

◆ Solve complex linear equations

A Signs of change?

To **add** an expression in brackets you can remove the brackets and then simplify.

Examples

$100 + (6 + 2)$	$100 + (6 - 2)$	$6 + (2p + 1)$	$(3n + 5) + (n - 6)$
$= 100 + 6 + 2$	$= 100 + 6 - 2$	$= 6 + 2p + 1$	$= 3n + 5 + n - 6$
$= 108$	$= 104$	$= 7 + 2p$	$= 4n - 1$

To **subtract** an expression in brackets you need to be careful with signs.

Examples

$100 - (6 + 2)$	$100 - (6 - 2)$	$6 - (2p + 1)$	$(3n + 5) - (n - 6)$
$= 100 - 6 - 2$	$= 100 - 6 + 2$	$= 6 - 2p - 1$	$= 3n + 5 - n + 6$
$= 92$	$= 96$	$= 5 - 2p$	$= 2n + 11$

A1 Simplify the following expressions.

(a) $6z + (2z - 3)$ (b) $(y + 5) + (2y - 1)$ (c) $(3x + 2) + (x - 6)$

(d) $(10 + w) + (5 - 2w)$ (e) $(2v - 1) + (5v - 3)$ (f) $(6 - u) + (2 - 3u)$

A2 Simplify the following expressions.

(a) $7t - (2t + 9)$ (b) $10 - (s - 1)$ (c) $3r - (5 + 2r)$

(d) $10q - (3 - 2q)$ (e) $(8p + 6) - (5 + 3p)$ (f) $(2n + 1) - (9 - 2n)$

A3 Find four pairs of equivalent expressions.

A $2 + (3a - 8)$

B $4a - (3a + 6)$

C $(2a + 1) + (a - 9)$

D $(6a - 8) + (2 - 5a)$

E $4a - (6 - 3a)$

F $(8a - 7) + (1 - a)$

G $2a - (8 - a)$

H $5a - (6 + 2a)$

A4 Simplify the following expressions.

(a) $(m + 3) + (5m - 9)$ (b) $10k - (3k - 8)$ (c) $(6j + 9) - (4j + 5)$

(d) $(7h - 3) + (5 - 2h)$ (e) $(7g - 3) - (5 - 2g)$ (f) $(f + 2) - (5 - f)$

(g) $(9 - 3e) - (7 + 8e)$ (h) $(5d - 6) - (3d - 5)$ (i) $(5c - 6) + (3c - 5)$

(j) $(10 - 5b) + (2 - 3b)$ (k) $(10 - 5a) - (2 - 3a)$ (l) $(6x + 3) - (7x + 1)$

A5 Solve the puzzle on sheet P13.

*A6 (a) Try some numbers for this puzzle and describe what happens.

(b) Copy and complete the algebra box to explain how the puzzle works.

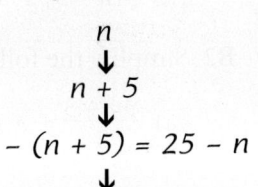

Puzzle

Think of a number

- Add 5

- Subtract **from** 30

- Add the number you first thought of

What is the result?

Algebra

n

↓

$n + 5$

↓

$30 - (n + 5) = 25 - n$

↓

*A7 For each puzzle below

(a) Try some numbers and describe what happens.

(b) Use algebra to explain how the puzzle works.

1 Think of a number

- Multiply by 3

- Add 9

- Subtract from 12

- Divide by 3

- Add the number you first thought of

What is the result?

2 Think of a number

- Subtract from 20

- Multiply by 2

- Subtract from 100

- Divide by 2

- Subtract 30

What is the result?

3 Think of a number

- Subtract 1

- Multiply by 4

- Subtract from 40

- Divide by 4

- Add the number you first thought of

What is the result?

B *Further simplifying*

It is often a good idea to multiply out any brackets and simplify divisions before sorting out any signs.

Examples

$$(3n + 5) - 2(n + 6)$$
$$= (3n + 5) - (2n + 12)$$
$$= 3n + 5 - 2n - 12$$
$$= n - 7$$

$$2(5n - 4) - 3(2n - 6)$$
$$= (10n - 8) - (6n - 18)$$
$$= 10n - 8 - 6n + 18$$
$$= 4n + 10$$

$$\frac{3n + 6}{3} + 2(n + 5)$$
$$= \frac{3n}{3} + \frac{6}{3} + (2n + 10)$$
$$= n + 2 + 2n + 10$$
$$= 3n + 12$$

B1 Simplify the following expressions.

(a) $6n + 7(n + 1)$ (b) $(3n + 5) + 2(2n - 1)$ (c) $2(n + 3) + 6(2n - 3)$

(d) $9(n - 3) + 5(3 - n)$ (e) $2(2n + 1) + 4(3 - n)$ (f) $3(n - 2) + 6(n + 1)$

B2 Simplify the following expressions.

(a) $\dfrac{9n + 3}{3} + 5(n + 2)$ (b) $\dfrac{10n + 8}{2} + 3(5n - 2)$ (c) $\dfrac{6n + 9}{3} + 2(5 - n)$

B3 Simplify the following expressions.

(a) $12 - 2(x + 3)$ (b) $(7x + 5) - 3(x + 1)$ (c) $5(3x + 3) - 4(2x - 3)$

(d) $7(x - 3) - 3(3 + x)$ (e) $10x - 3(2x - 5)$ (f) $4(2x + 3) - 6(2 - x)$

B4 Simplify the following expressions.

(a) $\dfrac{20x + 8}{4} - 3(2 + x)$ (b) $\dfrac{14x + 21}{7} - 3(6 - 2x)$ (c) $\dfrac{10x - 14}{2} - 2(5 - x)$

B5 Simplify the following expressions.

(a) $5p + 2(5 - 4p)$ (b) $(10p - 3) - 6(p + 2)$ (c) $2(5p + 3) - 9(p - 2)$

(d) $\frac{1}{4}(8p - 20) + 2(6 - p)$ (e) $\dfrac{45p + 40}{5} - 2(3 + 2p)$ (f) $\dfrac{12p + 36}{6} - 3(2 - 3p)$

*B6 Copy and complete each statement

(a) $2(3x + 1) + \blacksquare(x - \blacksquare) = 11x - 13$

(b) $\blacksquare - 3(2 - p) = 13p - 6$

(c) $2(3m - \blacksquare) - \blacksquare(2m - 5) = 13$

(d) $\dfrac{8h - 12}{\blacksquare} - 3(\blacksquare - h) = 5h - 6$

C Equations

C1 (a) Simplify the expression $5n + 3(n - 5)$.

    (b) Use the result of part (a) to solve the equation $5n + 3(n - 5) = 1$.

C2 (a) Simplify the expression $10 - (5 - c)$.

    (b) Use the result of part (a) to solve the equation $10 - (5 - c) = 20$.

C3 (a) Simplify the expression $12 - 3(1 + e)$.

    (b) Use the result of part (a) to solve the equation $12 - 3(e + 1) = 21$.

C4 Solve

    (a) $17 + 5(c - 4) = 47$ (b) $16 - 2(d + 3) = 7$ (c) $6f - 3(2 - f) = 12$

    (d) $5 - (g - 6) = 1$ (e) $h + 3(h - 1) = 3$ (f) $25 - 5(3k + 4) = 20$

    (g) $3m - 5(2 - m) = 10$ (h) $2(3 - 2n) + 5(n + 1) = 8$ (i) $3(p + 1) - 2(1 - 4p) = 34$

C5 Solve

    (a) $12 - (3 + q) = q - 1$ (b) $5r + 2(6 + r) = 3r + 8$

    (c) $10s - 3(2s - 1) = 6(s - 1)$ (d) $6(t - 2) - 5(1 - 3t) = 2(t + 1)$

Test yourself with these questions

T1 Simplify these expressions.

    (a) $10 - (4 + 5x)$ (b) $2(n + 3) + 5(n - 6)$ (c) $4(3m - 2) + 3(2m - 1)$

    (d) $\dfrac{4n + 16}{4} - 3n$ (e) $10k - 3(1 - 2k)$ (f) $5(3x - 4) - 7(2x - 5)$

T2 Solve these equations.

    (a) $6m + 2(m - 3) = 58$ (b) $2(c + 6) + 3(c - 1) = 4$

    (c) $n - 3(2 - n) = 4$ (d) $3(2k - 5) + 5(3k + 1) = 179$

T3 Solve these equations.

    (a) $5s + 2(8 - s) = 17(s - 4)$ (b) $2(x + 1) - 3(x - 5) = 5(x + 7)$

15 Distributions

You will revise

♦ how to find the mean, median, mode and range of sets of data

♦ how to put data into frequency tables

You will learn

♦ how to use stem and leaf tables

♦ how to use grouped frequency tables

♦ how to estimate the mean from a grouped frequency table

A Stem and leaf tables

Put your finger on the pulse

First aiders take a person's pulse from the carotid artery which is in the neck.

The pulse rate is recorded as the number of beats per minute (b.p.m.).

What is your pulse rate now?

Is your pulse faster at the beginning of the lesson or at the end?

Here are the pulse rates of a class of Year 10 students taken at the beginning of a lesson:

Pulses (beats per minute)   *77, 74, 85, 77, 72, 73, 55, 60, 91, 85, 80, 83, 68, 71, 60, 77, 77, 86, 97, 47, 71, 72, 63, 68, 84, 87, 77, 61*

A useful way to record this data is a stem and leaf table.

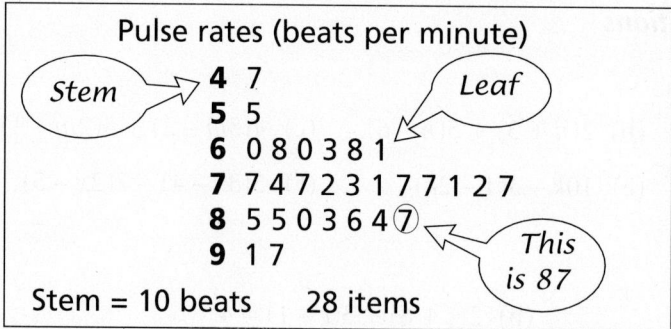

By putting the leaves in order in each row the table is even more useful.

Pulse rates (beats per minute)

4 7
5 5
6 0 0 1 3 8 8
7 1 1 2 2 3 ④ ⑦ 7 7 7 7
8 0 3 4 5 5 6 7
9 1 7

> Since there are 28 items of data the median is between the 14th and 15th. It is easy to count from the table that it is between 74 and 77. So the median is 75.5 b.p.m.

Can you find the range easily from this table?
Can you find the mode or modal group easily from this table?

A1 A class measured how long they could hold their breath.
Here are their results:

Breath holding (seconds): 40, 41, 69, 55, 48, 28, 18, 43, 45, 63, 50, 57, 80, 38, 14, 40, 40, 58, 60, 41, 65, 48, 41, 32

(a) Put this data into a stem and leaf table.
Make another copy of the table putting the data in order.

(b) Use your table to find the median time the students in this class could hold their breath for.

(c) Write down the range of the times.

A2 The data below shows the average male life expectancy of the main countries in Africa.

Life Expectancy (Years) in 2000

Algeria	68	Madagascar	57	South Africa	62
Angola	45	Mauritius	68	Sudan	54
Ethiopia	48	Nigeria	51	Tunisia	68
Ghana	56	Rwanda	41	Uganda	40
Kenya	52	Sierra Leone	36		

(a) Put this data into a stem and leaf table with the data in order.

This is the data for females in the same African countries.

Female life expectancy (years)
3 | 9
4 | 2 3 8
5 | 2 4 6 6
6 | 0 0 8
7 | 0 1 5
Stem = 10 years   14 items

(b) Compare the life expectancies of males and females in these African countries.

A3 This data shows the hours of sunshine recorded in London every day one September.

Sunshine (hrs) 10.1 9.6 9.6 6.8 7.4 10.0 11.8 10.6 8.9 5.8 0.0 8.7 9.4 5.5 6.1
0.1 8.1 1.7 8.3 10.2 7.8 9.1 0.1 0.1 3.6 3.4 0.1 0.7 0.2 9.3

  (a) Put this information into an ordered stem and leaf table with a stem of 1 hour units.

  (b) Use your table to find the median and range of the hours of sunshine in September.

A4 Stem and leaf tables can be useful for showing two sets of data side by side.
This data shows the pulse rates of a class at the beginning and end of a PE lesson.

Pulse rates (beats per minute)
Before PE   After PE

```
        8 7 | 4 |
    7 4 2 1 0 0 | 5 | 3 5
  8 8 6 5 5 3 1 | 6 | 0 3 6 8 9
      7 5 5 1 0 | 7 | 1 4 4 7 9 9
          6 2 1 | 8 | 3 3 6 6 7 8 9 9
            5 0 | 9 | 0 3 5 8
```

Stem = 10 beats   25 items

  (a) Find the median and range of the pulse rates,

    (i) before the PE lesson

    (ii) after the PE lesson

  (b) Use the table and your answer to (a) to write a brief statement about the pulse rates
  of the class before and after their PE lesson.

A5 This data shows the pulse rates of four different groups of people in Southbury.

A: Southbury Athletic Club	**B: Babies at Southbury Post Natal Clinic**	**C: Southbury Pensioners Club**
4 8 9	4	4 3 7
5 3 6 7	5	5 4 6
6 2 2 3 4 5 6 9	6	6 3 4 6
7 1 2 4 8	7 4 5 8	7 3 4 6 7
8 0 0 4	8 0 1 2 4 6 7 9	8 2 5 6
9	9 0 3 5 5 7 8	9 2 5
10	10 3 5 6 7 6	10 4
11	11 5	11
Stem.= 10bpm	Stem = 10bpm	Stem = 10bpm

Compare the pulse rates of the three groups
Calculate any figures which help you to make comparisons.

B Grouped frequencies

Two friends are playing 'Shove coin'.

The idea is to tap with the palm of their hand to get the coin as close to the target line as possible.

If the coin goes over the line, the player goes again. Each player records their shove by making a small mark on the next grid line between the front edge of the coin and the target line.

Each player has 50 shoves.

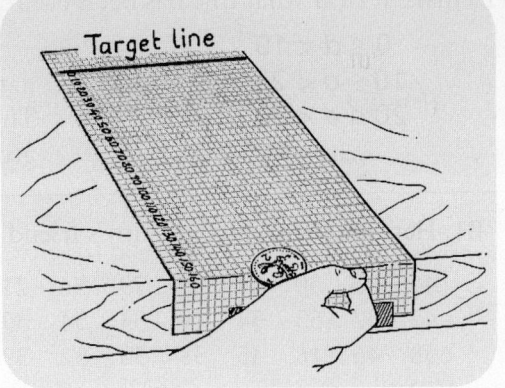

The results for Amy are shown here.

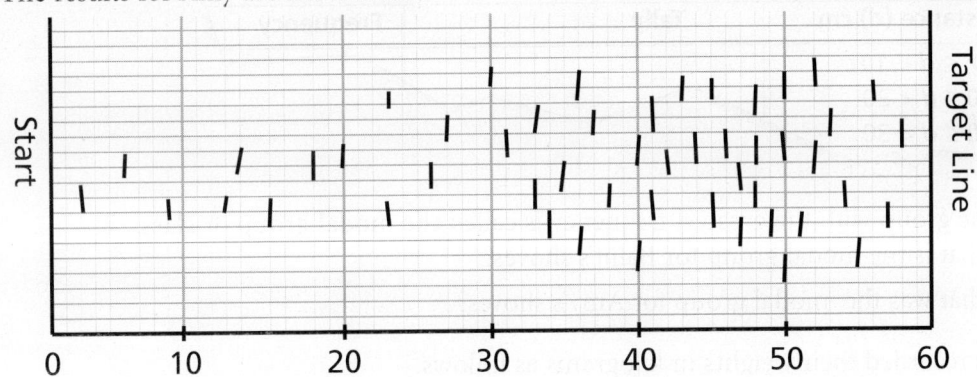

One way to record Amy's results is in a grouped frequency table.

Distance from start (cm)	Frequency
0 - 10	3
10 - 20	4
20 - 30	5
30 - 40	10
40 - 50	16
50 - 60	12
Total	**50**

- Which group did the 20 mark go into?
- Which group did the three 50 marks go into?

  How can you make it clear which group marks at the ends of groups go into?

It does not matter whether marks go in the group above or below provided the same rule is used with every group.

To make it clear what rule has been used here one way the **intervals** can be described is:

$0 \leq d < 10$

$10 \leq d < 20$ — *So 20 goes in this group* ... where d is the distance from the start.

$20 \leq d < 30$ The $\leq$ sign shows that all the numbers in that

.... group are greater than or equal to the first number.

B1 Here are the results for Amy's friend Baljit.

44	5	30	44	49	29	56	21	12	56	24	57	14	54	7	48	17	18
47	32	53	54	19	54	54	39	26	12	10	22	41	52	58	49	23	40
58	46	18	45	38	50	52	39	37	28	39	43	51	36				

(a) Copy and complete this table for Baljit's results.

Distance (d) cm	Tally	Frequency
$0 \leq d < 10$		
$10 \leq d < 20$		
$20 \leq d < 30$		

(b) The group with the greatest frequency is called the **modal group** or **class**. What is the modal group for Baljit's shoves?

(c) What was the modal group for Amy's shoves?

B2 A class recorded their weights in kilograms as follows:

Weight (kg)   46   45   52   61   57   47   61
52   47   47   42   59   51   35
48   62   62   47   52   39   72   69
57   43   50   38   61   47   54   40

The teacher asks the students to record this data in a grouped frequency table with these groupings.

(a) Which group will the weight 50 kg go into?

(b) Which group will the weight 40 kg go into?

(c) Copy and complete a table showing these students' weights using these groupings. Check that you have recorded the correct number of weights.

(d) Write down the modal group of weights.

(e) Use your table to copy and complete this frequency diagram for the weights of students.

Weight w (kg)	Frequency
$30 < w \leq 40$	
$40 < w \leq 50$	
$50 < w \leq 60$	
$60 < w \leq 70$	
$70 < w \leq 80$	
Total	

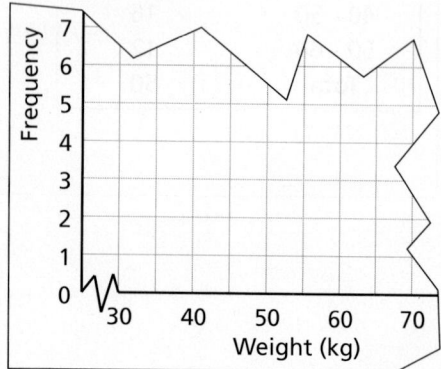

B3 The following is a record of the heights, in centimetres, of 40 guinea pigs.

21 22 11 16 22 13 11 25   9 17 21 24 27 25 12 14   8 12   6 17
23   7 12 26 14   8 12 26 17 19 23 29 21 19 26 26 18 21 13   9

(a) Copy and complete this frequency table.

Height (h) cm	Tally	Frequency
$5 \le h < 10$		
$10 \le h < 15$		
$15 \le h < 20$		
$20 \le h < 25$		
$25 \le h < 30$		

(b) Draw a frequency diagram for this information on squared paper.

(c) How many guinea pigs were under 15 cm in height?

(d) Write down the modal class interval of the heights.

[Edexcel 1998]

B4 The maximum daily temperature for the month of October was recorded at a Sussex weather station as follows:

Max. Temp. (°C) 16.7 13.9 15.4 14.6 14.1 14.2 12.5 14.6 12.2 12.9 12.2 13.1 14.7
                12.7 13.9 12.7 13.9 14.4 13.9 14.5 14.7 14.1 14.7 12.9 14.4 13.1
                13.6 14.6 12.8 13.3 12.4

(a) Copy and complete this grouped frequency table to record the temperatures.

Max. Temp. (°C)	Frequency
$12.0 \le t < 13.0$	
$13.0 \le t < 14.0$	
$14.0 \le t < 15.0$	
$15.0 \le t < 16.0$	
$16.0 \le t < 17.0$	
Total	

(b) What is the modal group of temperatures in Sussex during October?

(c) On how many days in October did the temperature not reach 14°C?

ℂ *In the interval*

This data shows the heights in cm of 44 male African elephants:

272 273 287   84   95 153 165 161 168 257 262 293 194 193 204 218 218
227 186 181 182 224 231 236 237 256 260 238 235 247 200 207 201 215
245 290 317 108 124 270 287 121 135 142

These frequency tables and diagrams show this data grouped in different intervals.

Height h (cm)	Frequency
$80 \le h < 100$	2
$100 \le h < 120$	1
$120 \le h < 140$	3
$140 \le h < 160$	2
$160 \le h < 180$	3
$180 \le h < 200$	5
$200 \le h < 220$	7
$220 \le h < 240$	7
$240 \le h < 260$	4
$260 \le h < 280$	5
$280 \le h < 300$	4
$300 \le h < 320$	1
Total	44

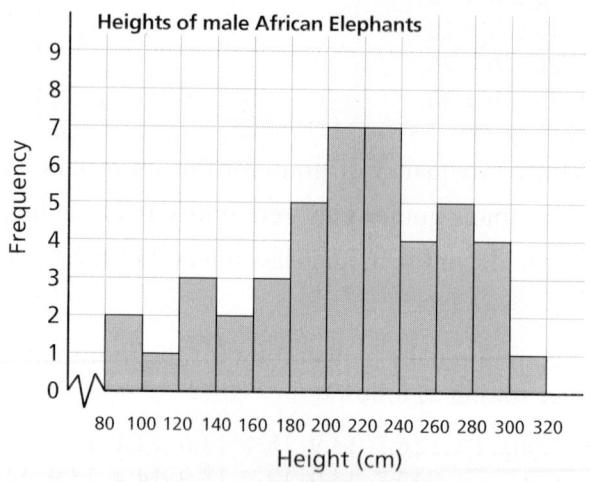

Height h (cm)	Frequency
$50 \le h < 100$	2
$100 \le h < 150$	5
$150 \le h < 200$	9
$200 \le h < 250$	16
$250 \le h < 300$	11
$300 \le h < 350$	1
Total	44

Height h (cm)	Frequency
$0 \le h < 100$	2
$100 \le h < 200$	14
$200 \le h < 300$	27
$300 \le h < 400$	1
Total	44

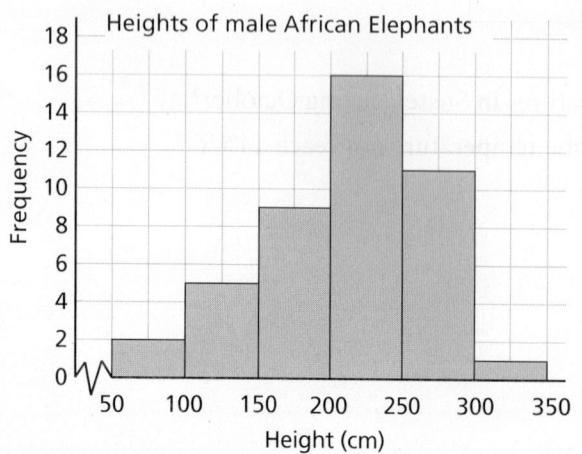

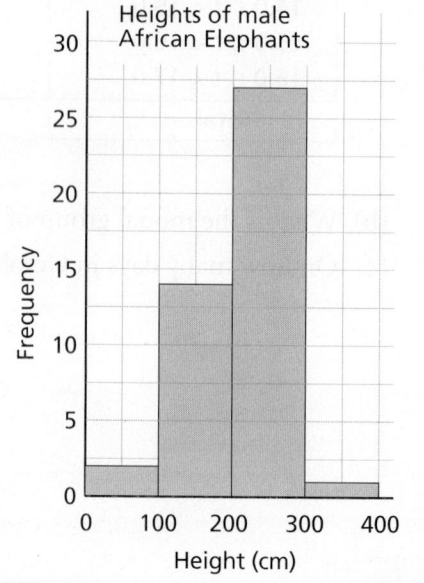

- Which way of grouping gives the clearest picture of the pattern of the data?
- Which way gives the most useful value for a modal class?
- Would using intervals of 5 cm give a clearer picture?
  Would intervals of 200 cm give you more information?
- Would using intervals of 35 cm be a useful grouping?

C1 This data shows the footlength (f) in centimetres of the 44 male African elephants.

49 49 51 17 18 28 30 30 31 46 47 52 35 35 37 40 40 41 34 33 33 41
42 43 43 46 47 43 42 45 36 38 37 39 44 52 57 21 23 48 51 23 25 26

(a) Record this data in a frequency table with intervals
$15 \leq f < 20$, $20 \leq f < 25$, $25 \leq f < 30$ ...
What is the modal group?

(b) Write out another table with intervals $10 \leq f < 20$, $20 \leq f < 30$, $30 \leq f < 40$ ...
What is the modal group now?

(c) Which of these two sets of intervals gives the clearest picture of the data?

C2 For each of these sets of data decide on a suitable set of intervals.
Put the data into a frequency table using these intervals.
Draw a frequency diagram from your table.

(a) | Reaction times (thousandths of a second) |
    | 20 18 18 13 18 15 16 17 17 18 22 27 17 15 11 12 12 12 13 |
    | 17 18 21 19 15 16 10 11 16 15 20 16 22 15 14 14 |

(b) | *Weights (kg)* |
    | *62.5 63.1 62.9 63.8 62.4 65.7 65. 3 66.9 64.7 66.1 65.0 65.2 67.6 66.3* |
    | *67.3 68.1 69.0 68.4 69.6 68.9 68.5 69.2 70.5 71.3 70.4 71.8 70.0 70.8* |
    | *70.3 70.7 71.2 70.1 72.4 73.6 72.7 72.9 74.5 75.1* |

Open ended

This data records the time in seconds between eruptions at the Kiama Blowhole
near Sydney, Australia.

83 51 87 60 28 95   8 27 15 10 18 16 29 54 91   8 17 55 10 35 47 77
36 17 21 36 18 40 10   7 34 27 28 56   8 25 68 146 89 18 73 69   9 37
10 82 29   8 60 61 61 18 169 25   8 26 11 83 11 42 17 14   9 12

- What would be a suitable set of intervals to record this data?

When there are just a few 'extreme' pieces
of data open ended groups are often used.
The data above could be recorded in these groups:

Can you easily draw a frequency diagram
from this table?

Time (s) secs
$0 \leq s < 20$
$20 \leq s < 40$
$40 \leq s < 60$
$60 \leq s < 80$
$80 \leq s < 100$
$100 \leq s$

This group includes any times greater than 100 secs

Ⓓ *Estimating means*

This frequency diagram and table summarises the amount collected in a day by people collecting for a charity.

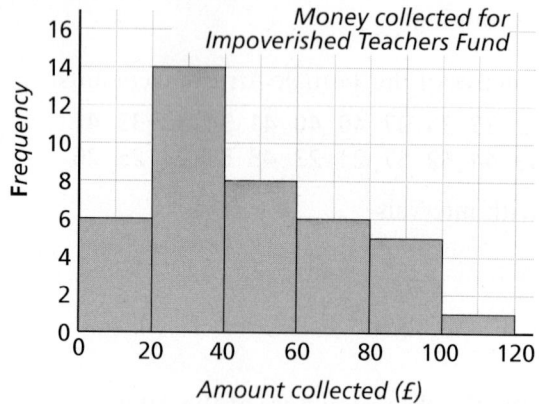

Money (m) £	Frequency
$0 \le m < 20$	6
$20 \le m < 40$	14
$40 \le m < 60$	8
$60 \le m < 80$	6
$80 \le m < 100$	5
$100 \le m < 120$	1
Total	40

To find the mean amount collected per person the total amount collected is needed.

The 6 people in the $0 \le m < 20$ interval all collected between 0 and £20.
As the average amount collected by this group was probably around £10 a reasonable estimate of the amount collected by this is group is 6 × £10 = £60.

A reasonable estimate of the total amount collected would be
$$(6 \times 10) + (14 \times 30) + (8 \times 50) + (6 \times 70) + (5 \times 90) + (1 \times 110) = £1860$$

So an estimate of the mean amount collected per person is 1860 ÷ 40 = £46.50

Since this is only an estimate it would be sensible to quote the mean as £47 or even £50.

D1 A local police force want to estimate the mean speed of cars along a particular stretch of road which has a 40 m.p.h. speed limit.
This table shows the speeds recorded one morning and some unfinished working to estimate the mean speed.

Speed s (m.p.h.)	Frequency	Mid-interval value	Group total estimate
$20 < s \le 30$	7	25	7 × 25 = 175
$30 < s \le 40$	21	35	21 × 35 = 735
$40 < s \le 50$	8		
$50 < s \le 60$	3		
$60 < s \le 70$	1		
Total	40		

(a) How many people were breaking the speed limit?

(b) Copy and complete the table above.

(c) Use your table to estimate the mean speed of the cars.

D2 This bar chart gives information about the temperatures of a group of patients with a particular illness.

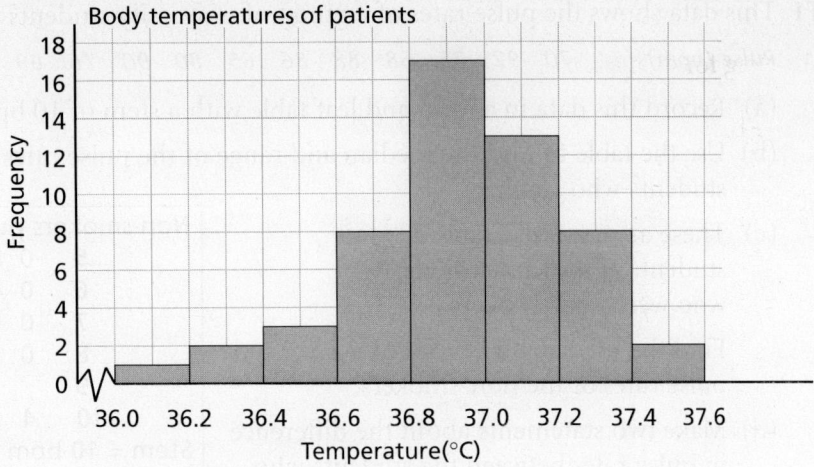

Body temperatures of patients

(a) How many patients were in the group altogether?

(b) Use the chart to write out a grouped frequency table with groups $36.0 < t \le 36.2$, ...

(c) Use your table to find an estimate of the mean temperature of these patients.

D3 On holiday Val records the length of time people stay in the pool. The results are shown in the table.

Calculate an estimate of the mean time spent in the pool.

Give your answer to an appropriate degree of accuracy.

Time, t (mins)	Number of people
$0 < t \le 10$	4
$10 < t \le 20$	7
$20 < t \le 30$	3
$30 < t \le 40$	2
	16

[AQA 1999]

D4 This grouped frequency table shows the marks in a test.

(a) Meg says that the mid-interval value for the first group is 2.5. Noel says that it must be 3. Which is correct and why?

(b) Copy the table and add appropriate columns to find an estimate of the mean mark in the test.

Mark	Frequency
1 - 5	2
6 - 10	5
11 - 15	19
16 - 20	4
21 - 25	3

Test yourself with these questions

T1 This data shows the pulse rates of a group of university students who smoked.

Pulse (bpm)　　70　92　75　68　88　86　65　80　90　76　69　83　70　76　88

(a) Record this data in a stem and leaf table with a stem of 10 bpm.

(b) Use the table to find the median and range of the pulse rates of these students who smoked.

(c) These are the pulse rates of some students at the same university who were non-smokers.

Find the median and range of the pulse rates of the non-smokers.

Non-smokers pulse rates	
5	0 8 9
6	0 4 4 5 6 6 6 8 8
7	0 1 1 8
8	0 1 6 8 8
9	
10	4

Stem = 10 bpm

(d) Make two statements about the difference in pulse rates between the students who smoked and those who were non-smokers.

T2 A set of 25 times in seconds is recorded.

12.9　10.0　4.2　16.0　5.6　18.1　8.3　14.0　11.5　21.7　22.2　6.0　13.6
3.1　11.5　10.8　15.7　3.7　9.4　8.0　6.4　17.0　7.3　12.8　13.5

(a) Copy and complete the table below, using intervals of 5 seconds.

Time (t) seconds	Tally	Frequency
$0 \leq t < 5$		

(b) Write down the modal class interval.

[Edexcel]

T3 Draw a frequency diagram from the table in T2.

T4 Some women walked one mile. The time taken by each was recorded. The results are as follows.

Time *t* minutes	$12 \leq t < 16$	$16 \leq t < 20$	$20 \leq t < 24$	$24 \leq t < 28$	$28 \leq t < 32$
Number of women	1	9	43	22	5

(a) What is the modal class of the time taken?

(b) Calculate an estimate of the mean time taken.

One mile is approximately 1.6 km.

(c) Use the data to calculate an estimate of the mean time taken by these women to walk one kilometre.

[AQA 1998]

16 Volume and surface area

What you should know

◆ how to find the volume of a cuboid

What you should learn

◆ how to find the volume of a prism
◆ how to find the surface area of 3D shapes
◆ how to change between metric units in volume and area
◆ how to use density

A Volumes of cuboids

Volumes are measured in cubic centimetres (cm³)
or for larger volumes cubic metres (m³).

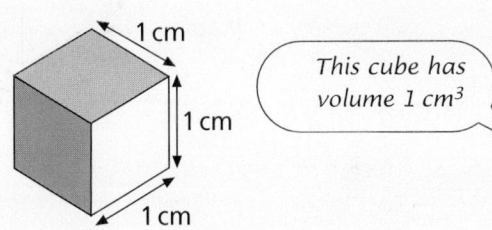

This cube has volume 1 cm³

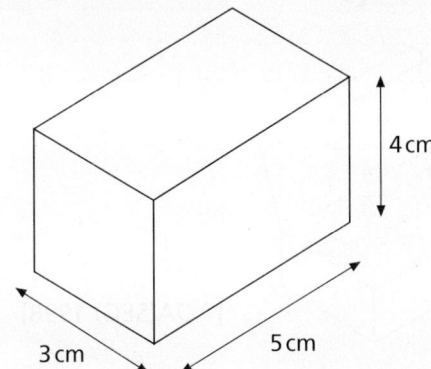

To find the volume of a cuboid use the formula

Volume = length × width × height

This cuboid has volume
$5 \times 3 \times 4 = 60 \text{ cm}^3$

A1 Find the volume of these cuboids.

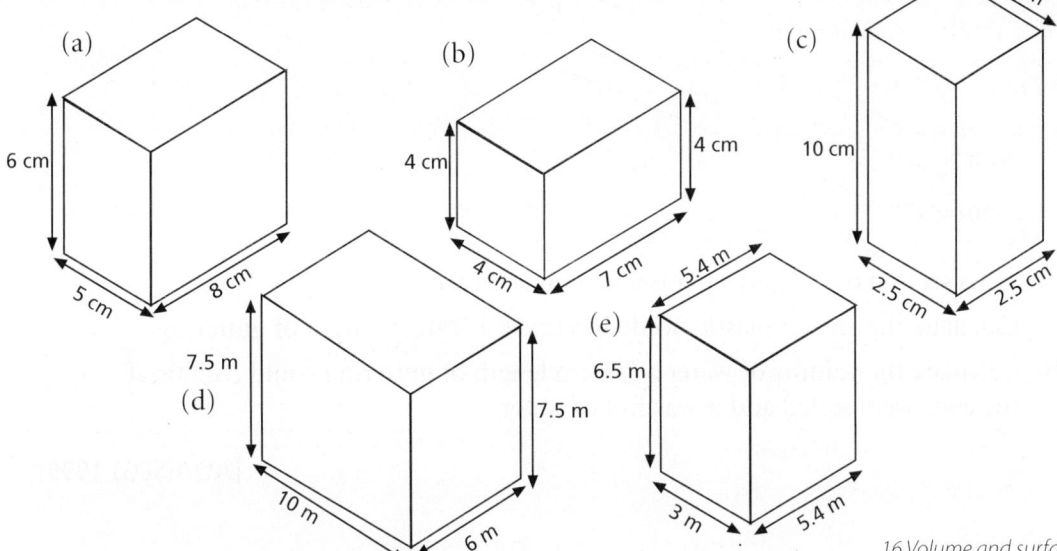

(a) 6 cm, 5 cm, 8 cm

(b) 4 cm, 4 cm, 7 cm, 4 cm

(c) 2.5 cm, 10 cm, 2.5 cm, 2.5 cm

(d) 7.5 m, 10 m, 6 m

(e) 5.4 m, 6.5 m, 7.5 m, 3 m, 5.4 m

A2 These cuboids all have the same volume.
Find the missing measurements. (They are not drawn to scale.)

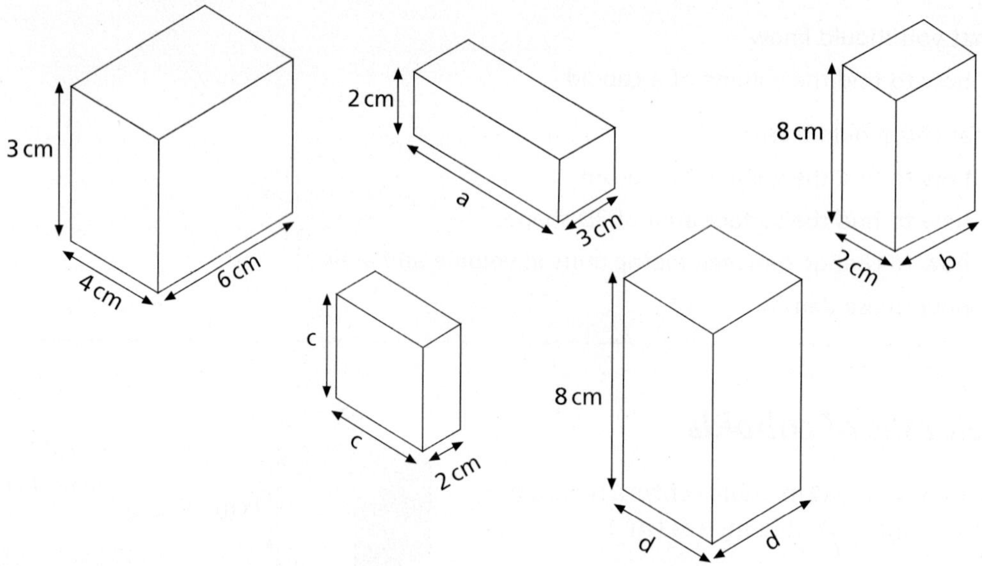

3 cm
4 cm
6 cm

2 cm
a
3 cm

8 cm
2 cm
b

c
c
2 cm

8 cm
d
d

A3 The diagram shows a cuboid which is just big enough to hold six tennis balls.

Each tennis ball has a diameter of 6.8 cm.

Calculate the volume of the cuboid.

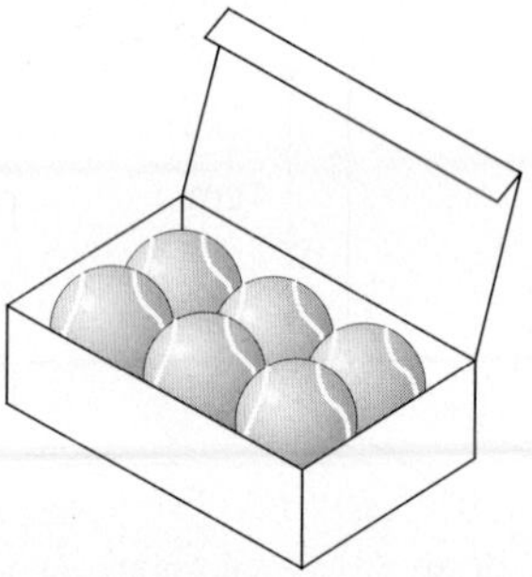

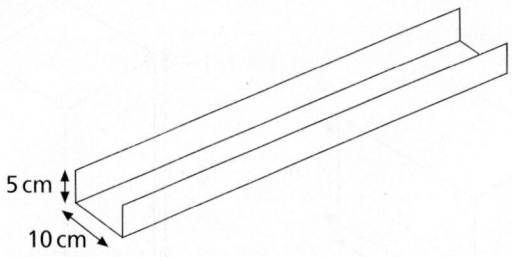

[AQA(SEG) 1998]

A4 This diagram shows a length of plastic guttering.

5 cm
10 cm

The cross section of the guttering is a rectangle measuring 10 cm by 5 cm.

(a) Calculate the area of plastic needed to make a 200 cm length of guttering.

(b) Calculate the volume of water a 200 cm length of guttering could contain if the ends were sealed and it was full of water.

[AQA(SEG) 1999]

B *Prisms*

This shape is made up of 20 cuboids of size 1 cm by 1 cm by 6 cm.

What is its volume?

The shapes below are all made from pieces 6 cm long. What are the volumes of these shapes?

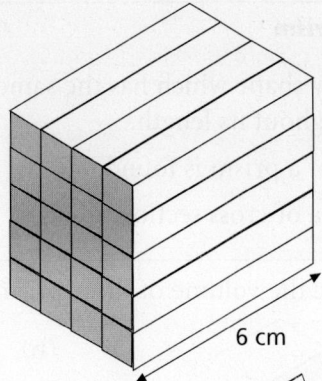

6 cm

A
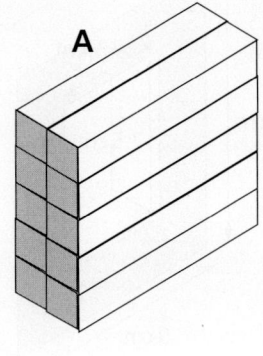

B
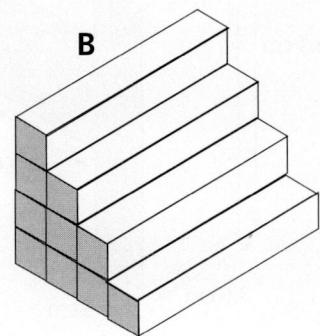

C
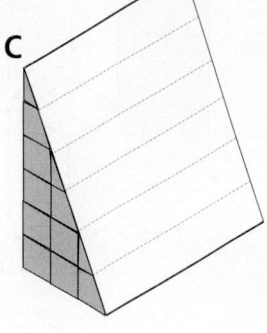

D
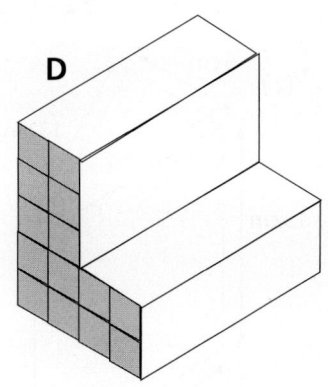

E
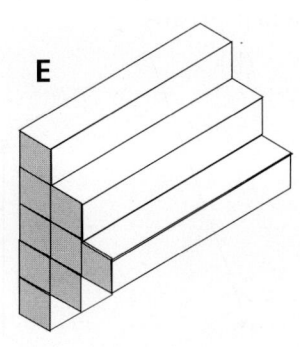

F
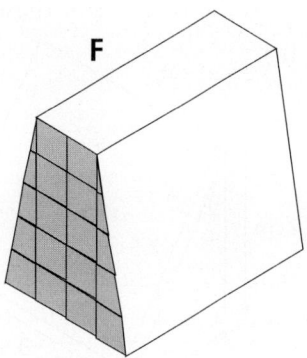

B1 Find the volume of these prisms

(a)
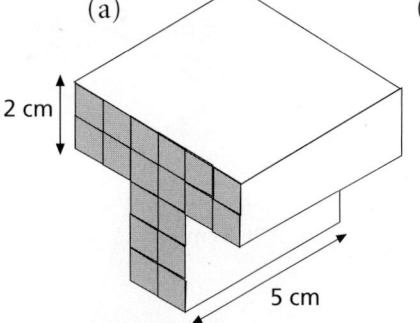

2 cm

5 cm

(b) (c)
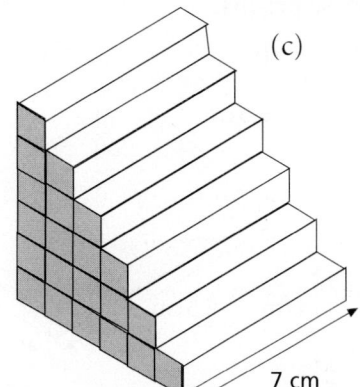

7 cm

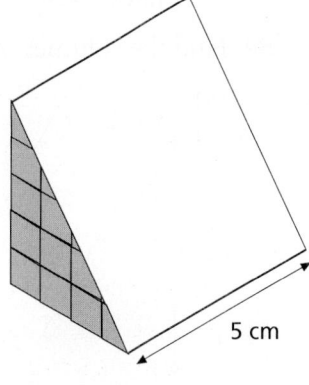

5 cm

Volume of a prism

A prism is any shape which has the same cross section throughout its length.

The volume of a prism is found by

Volume = area of cross section × length

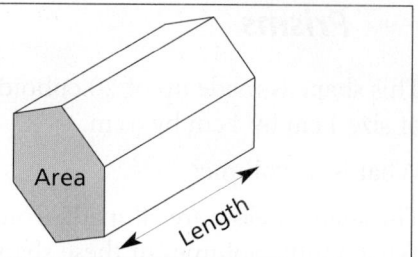

B2 Calculate the volume of these prisms. (Sketching the cross section may help.)

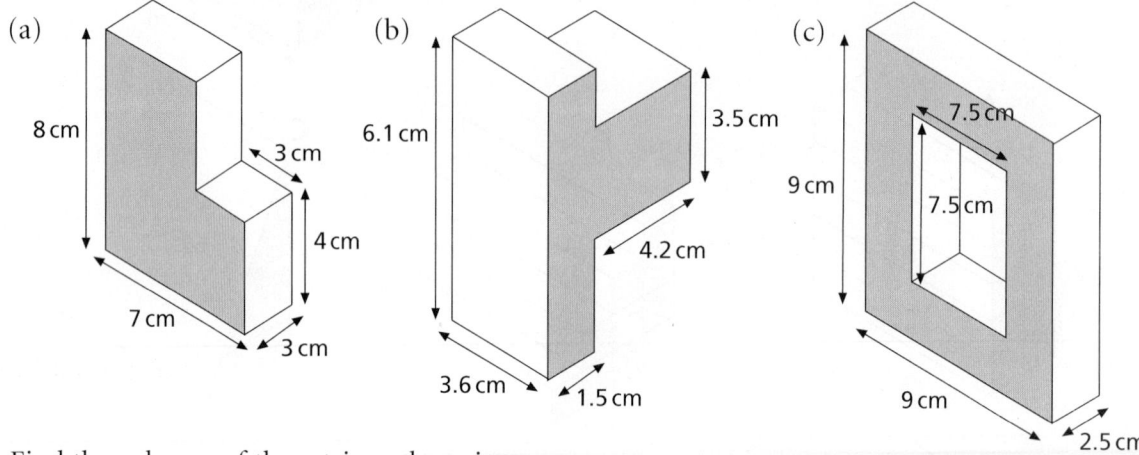

B3 Find the volumes of these triangular prisms.

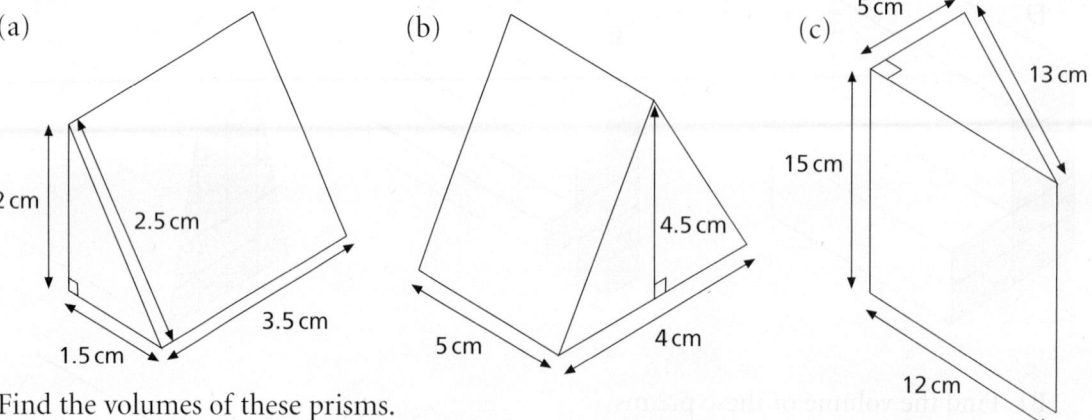

B4 Find the volumes of these prisms.

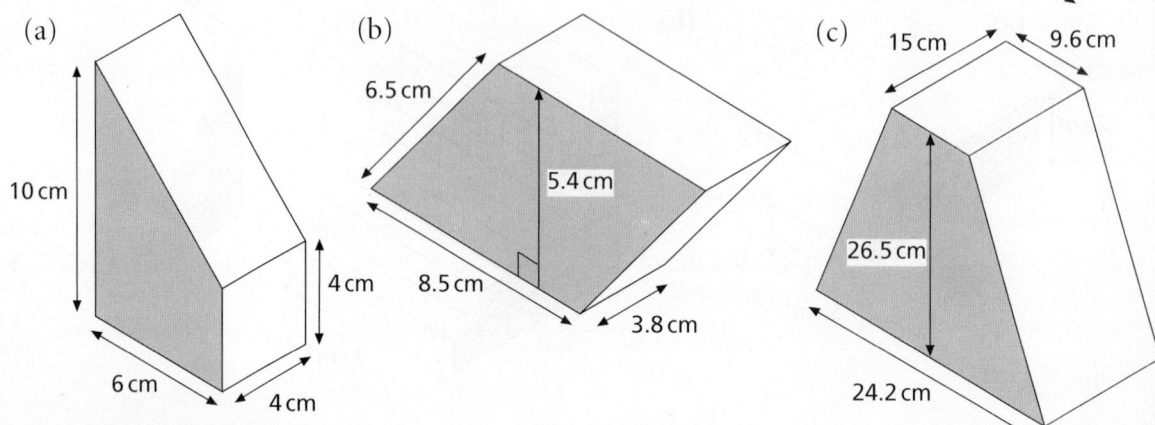

B5 A rubbish skip is made in the shape of a prism whose uniform cross section is a trapezium.

Find the volume of this skip.

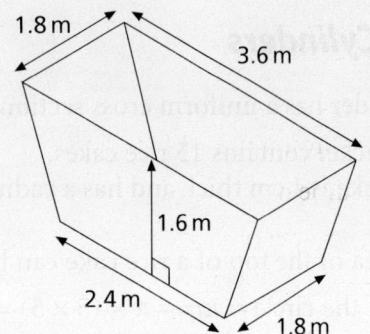

B6 Ice cream is sold in a box that is the shape of a prism.
The ends are parallelograms.
The size of the parallelograms is shown in the diagram.
The length of the prism is 12 cm.

Calculate the volume of the ice cream in the box.

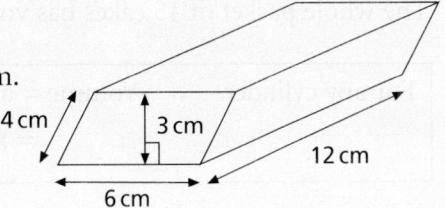

[AQA(NEAB) 1998]

B7 This triangular prism has a right-angled triangle cross section.
The volume of the prism is 66 cm³.

Find the length L.

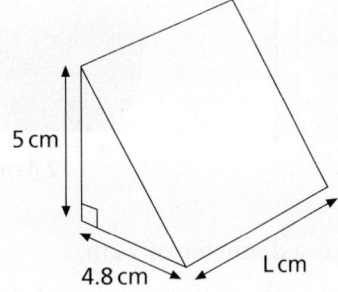

B8 A building site has area 150 m².
The topsoil on the site is 0.8 m deep throughout.

What is the volume of topsoil on the site?

B9 A block of glass is a cube measuring 15 cm on each side.
The glass is melted into a sheet which is 0.5 cm thick.

What will be the area of the sheet of glass?

B10 A concrete base is needed to cover a rectangular area 5 m by 3.6 m.
A cement mixer contains 4.5 m³ of concrete.

How thick will the concrete be if it is spread evenly over the area?

ℂ *Cylinders*

A cylinder has a uniform cross section which is a circle.

This packet contains 15 rice cakes.
Each cake is 1 cm thick and has a radius of 5 cm.

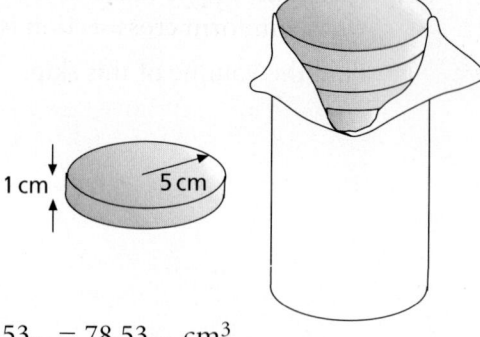

The area of the top of a rice cake can be found by

Area of the circle $= \pi r^2 = \pi \times (5 \times 5) = 78.5398\ldots\ldots$ cm^2.

Since the rice cakes are 1 cm thick they have volume $1 \times 78.53\ldots = 78.53\ldots$ cm^3.
The whole packet of 15 cakes has volume $15 \times 78.53\ldots = 1178.097\ldots = 1178.1$ cm^3 (1 d.p.).

For any cylinder $\qquad$ Volume = area of the circle $\times$ height $\qquad\qquad\qquad = \pi r^2 h$

C1 Find the volumes of these cylinders to 1 decimal place

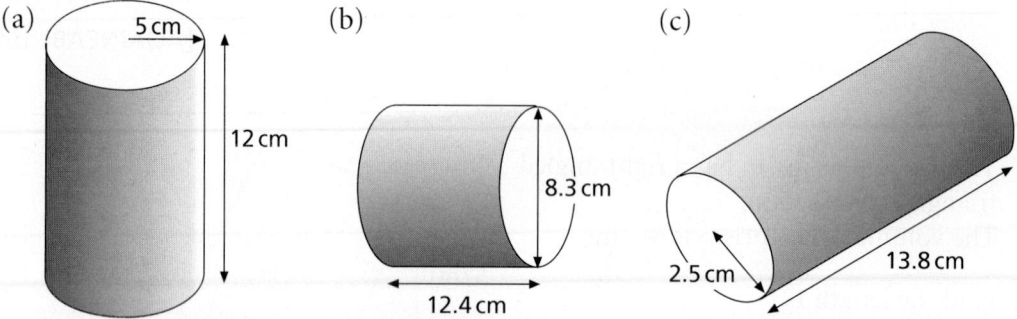

(a) 5 cm, 12 cm

(b) 8.3 cm, 12.4 cm

(c) 2.5 cm, 13.8 cm

C2 (a) A circle has a diameter of 7 cm.

    (i) Calculate the circumference of this circle.

    (ii) Calculate the area of this circle.

  (b) A plastic beaker has a height of 10 cm
     and a circular base of diameter 7 cm.
     Calculate the volume of the beaker.

]

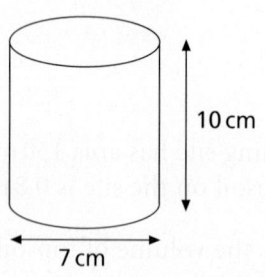

10 cm, 7 cm

[AQA(NEAB) 1997]

C3 Becca buys a large tin of coffee.
    It is a cylinder with radius 7.5 cm.
    It is filled to a depth of 14 cm.

    Work out the volume of coffee in the tin.
    State clearly the units of your answer.

[OCR]

Appropriate accuracy

A calculator will often give an answer with quite a few figures after the decimal point.

For example a factory wants to make a cylindrical can with volume $400\,cm^3$.
The radius of the can is to be 3.2 cm and they want to know how high the can must be.

The height is $400 \div (\pi \times 3.2^2) = 12.4339799291$ cm (from a calculator)

This is clearly far too accurate for practical purposes.
A suitable answer is   *Height = 12.433 … = 12.4 cm (to one decimal place)*

It has been rounded to one decimal place because the radius was given to one decimal place.
This appears to be the degree of accuracy that the factory uses.

However ... in November 2000 the Gateshead Millenium Bridge was put into place.
Although the bridge was made 126 m long it had to be made accurate to
within 3 mm if it was to fit into place when lowered in.

C4 Ranjit has made a new circular fish pond in his garden.
The radius of the pond is 1.5 m.

   (a) Calculate the circumference of the pond.

   (b) The sides of the pond are vertical.
   The water in the pond is 0.7 m deep.
   Calculate the volume of the water in the pond.
   Give your units in the answer.   [OCR]

C5 A gardener keeps rainwater in a cylindrical butt 1.5 m high with radius 50 cm.

   (a) What volume of water does the butt contain in cm^3 when it is full?

   (b) A litre is $1000\,cm^3$.
   How many 5 litre watering cans could the gardener fill from a full butt?

C6 Water is poured from a kettle containing $\frac{1}{2}$ a litre($500\,cm^3$) into a cup.
The cup has a radius of 4 cm.
How high up will the water in the cup come?

C7 A cylinder is 20 cm high and holds $1000\,cm^3$ of water.
Find the radius of the cylinder.

Capacity

On bottles and cans measurements are often given in millilitres (ml).
A millilitre is one thousandth of a litre and is a liquid measure the same as $1\,cm^3$.

Carefully measure some different size cans in centimetres.
Calculate the volume of these cans.

How do these volumes compare with the amount they are supposed to contain?

D Surface area

The surface area of an object is the total area of all its faces.
The net of an object clearly shows the areas that have to be calculated.

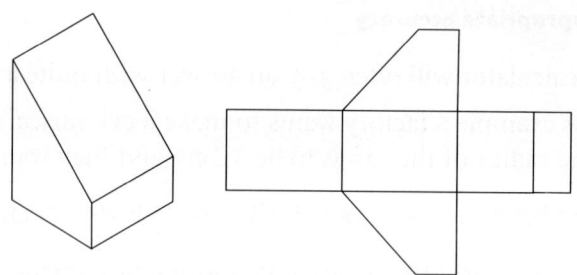

D1 For each of the objects below

(i) Sketch the net adding any measurements you know.
(ii) Find the surface area of the object.

(a)

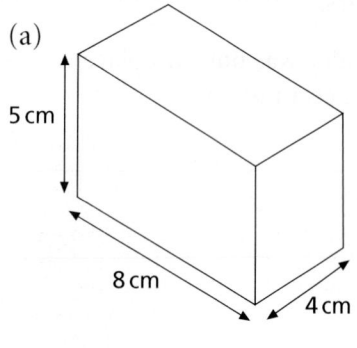

(b)

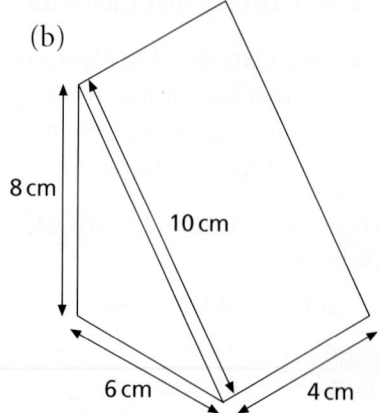

(c)

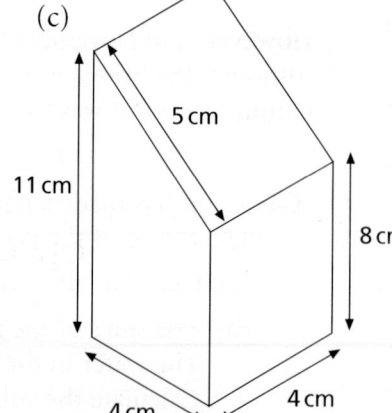

D2 (a) What is the circumference of the circle at each end of this cylinder?

(b) Find the area of one of the circular ends.

(c) Sketch the curved surface of the cylinder laid flat.
Add the dimensions of this shape.
Find the area of the curved surface of this cylinder.

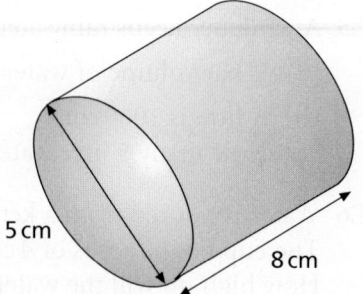

Carrying the can

A cylindrical can with a volume of 1000 cm³ (one litre) has radius 5 cm.
How high is it?
What would be the total surface area of material used to make this can?

Find the heights of 1 litre cans with radius (i) 3 cm   (ii) 4 cm   (iii) 6 cm   (iv) 7 cm
Find the surface area of each of these cans.

What radius should a 1 litre can be to use the least amount of material?

You may find a spreadsheet useful.

E *Density*

Watching your weights

Paul's grandfather asks him this question.
Do you know the answer?

> *Which weighs more, a tonne of feathers or a tonne of stones?*

The stones clearly have a much greater **density** than the feathers.
Density is usually given as as g/cm³. (Lead has density 11.4 g/cm³)

Weigh some solid objects you can easily find the
volume of, such as a brick or a block of polysterene.
Find their density.

$$\text{Density} = \frac{\text{Weight}}{\text{Volume}}$$

E1 Find the density of these objects

(a)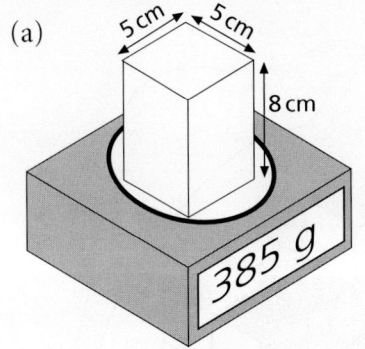

5 cm 5 cm 8 cm 385 g

(b)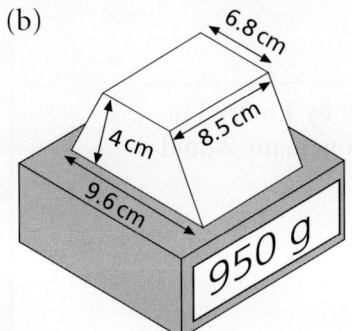

6.8 cm 8.5 cm 4 cm 9.6 cm 950 g

(c)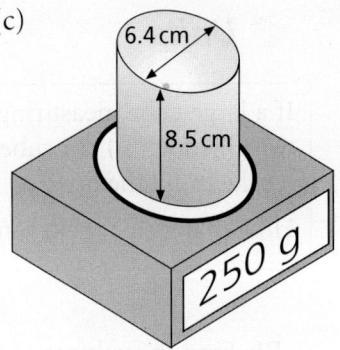

6.4 cm 8.5 cm 250 g

E2 Use the table on the right to work out the weight of these objects.

(a) A stone with volume 92 cm³

(b) A wooden cuboid 22 cm by 18 cm by 11 cm

(c) The water in a cylindrical container with
radius 40 cm and height 140 cm.

Material	Density
Wood	0.7 g/cm³
Stone	3.0 g/cm³
Water	1.0 g/cm³

E3 Density can help identify metals.

One of these ingots below is platinum, one is
gold, one is silver and one is fake gold.

Work out the density of each ingot and use this
table to say what each ingot is made of.

Material	Density
Platinum	21.5 g/cm³
Gold	19.3 g/cm³
Silver	10.5 g/cm³
Fake gold	9.8 g/cm³

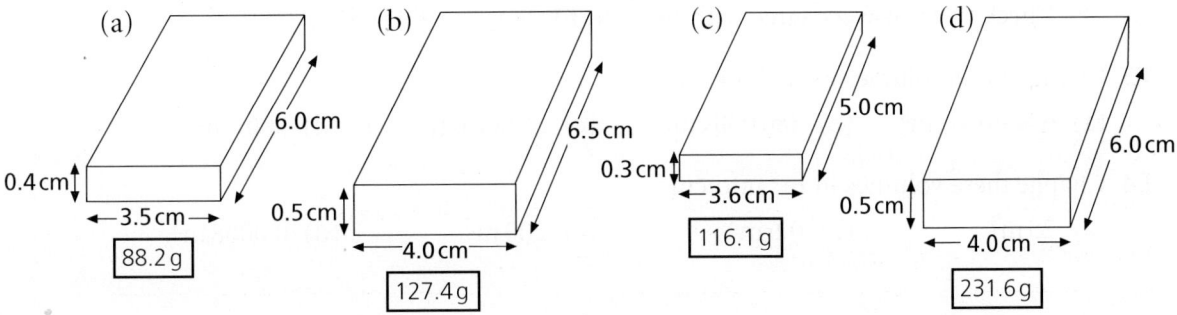

(a) 6.0 cm 0.4 cm 3.5 cm 88.2 g

(b) 6.5 cm 0.5 cm 4.0 cm 127.4 g

(c) 5.0 cm 0.3 cm 3.6 cm 116.1 g

(d) 6.0 cm 0.5 cm 4.0 cm 231.6 g

E4 This diagram shows a prism.
The cross section of the prism is a trapezium.
The lengths of the parallel sides of the trapezium
are 8 cm and 6 cm.
The distance between the parallel sides of the
trapezium is 5 cm.
The length of the prism is 20 cm.

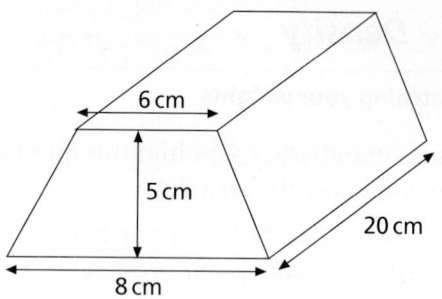

(a) Work out the volume of the prism.

The prism is made out of gold.
Gold has a density of 19.3 grams per cm³.

(b) Work out the mass of the prism.
Give your answer in kilograms.

[Edexcel]

F Units

If a large cube measuring 1 m by 1 m by 1 m
was filled with 1 cm cubes, how many would
it take to fill the box?

How many cm³ make a m³?

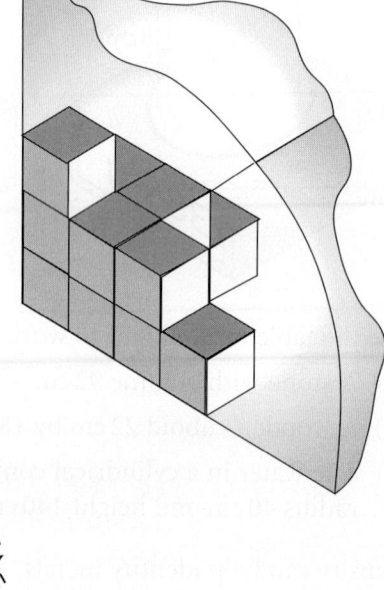

F1 Find the volume of these shapes in cm³.

(a)

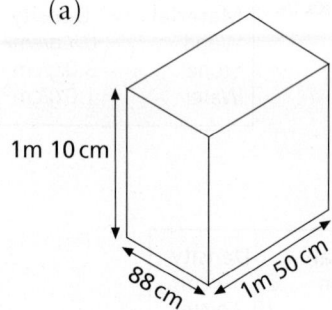

(b)

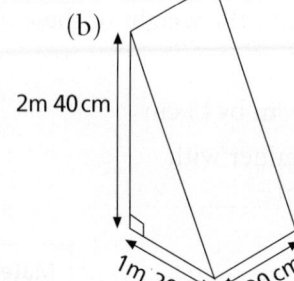

F2 (a) Sketch the shapes in F1 and change their measurements into metres.

(b) Find the volume of the shapes in m³.
Check your answers agree with those in F1.

F3 Change these volumes in cm³ into m³.

(a) 6 500 000 cm³   (b) 480 000 cm³   (c) 50 000 cm³   (d) 500 cm³

F4 Change these volumes in m³ into cm³ .

(a) 25 m³   (b) 0.6 m³   (c) 200 m³   (d) 0.0008 m³.

Test yourself with these questions

T1 Find the volume of these objects.

(a) a cuboid

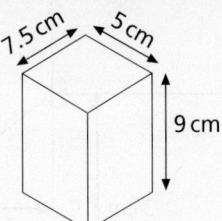

7.5 cm 5 cm 9 cm

(b) a triangular prism

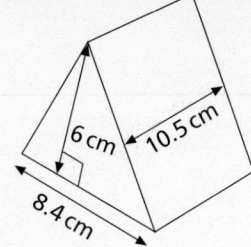

6 cm 10.5 cm 8.4 cm

(c) a cylinder

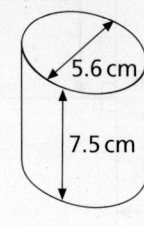

5.6 cm 7.5 cm

T2 *Chocomint* is made in blocks as shown in the diagram.

The block is a prism and its cross-section is a trapezium.

(a) Calculate the volume of a block of *Chocomint*.

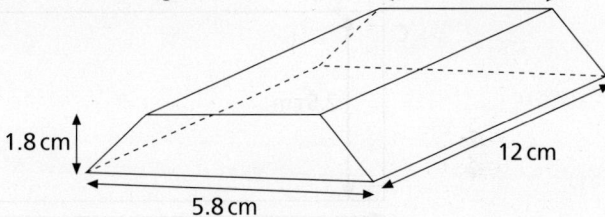

3.4 cm 1.8 cm 5.8 cm 12 cm

Chocorock is made in cylindrical sticks as shown in the diagram

The radius of the circular end is 1.2 cm.
The volume of the cylinder is 50 cm³.

1.2 cm

(b) Find the length of a stick of *Chocorock*.
Give your answer to a sensible degree of accuracy.

[OCR]

T3 A skip is in the shape of a prism with cross-section ABCD.
AD = 2.3 m, DC = 1.3 m and BC = 1.7 m.
The width of the skip is 1.5 m.

(a) Calculate the area of the shape ABCD.

(b) Calculate the volume of the skip.

The weight of an empty skip is 650 kg.
The skip is full to the top with sand.
1 m³ of sand weighs 4300 kg.

(c) Calculate the total weight of skip and sand.

2.3 m A D 1.3 m 1.5 m B 1.7 m C

[Edexcel]

T4 This net will fold to make a three dimensional shape.

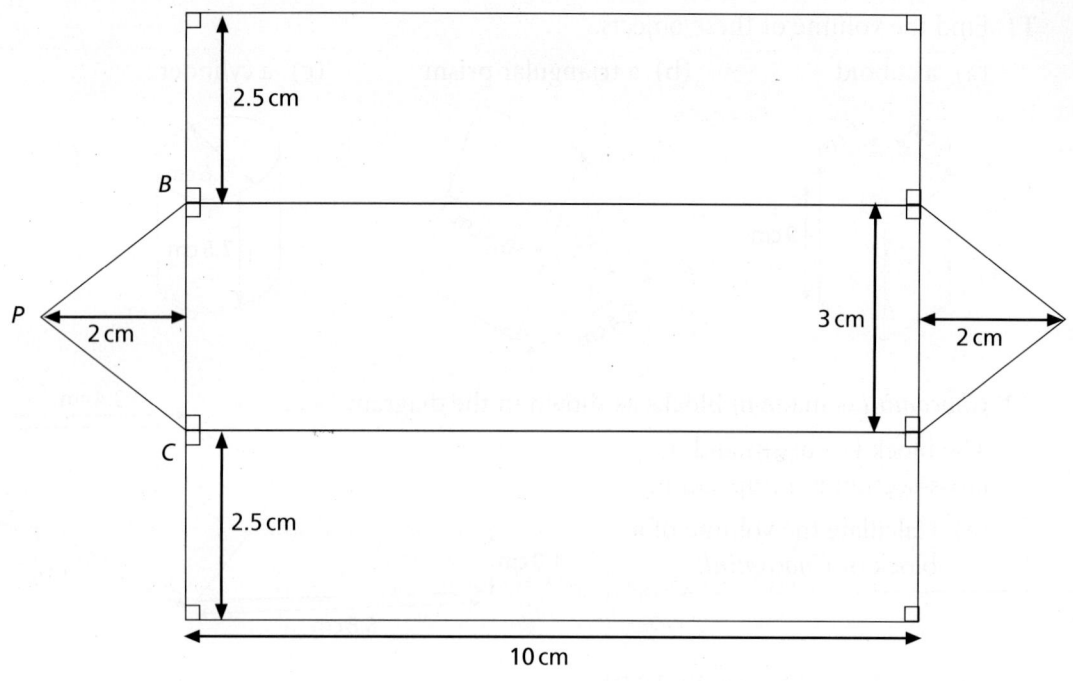

(a) Measure the size of the angle BPC.

(b) Calculate the area of the net.
(Remember to state the units in your answer)

(c) (i) Draw a sketch of the shape it makes when folded.

    (ii) What is the mathematical name of this shape?

    (iii) Calculate the volume of this shape.
       (Remember to state the units in your answer.)

[AQA 1997]

Review 3

1 Simplify the following expressions.

   (a) $4w - (9 + 2w)$ (b) $8b - 7 - (3 + 2b)$ (c) $(5x + 3) - (6x + 1)$

   (d) $6q + 3(4 - 5q)$ (e) $11x - 4(3x - 2)$ (f) $2(6a + 4) - 10(2a - 3)$

2 (a) Write a formula for the perimeter of this rectangle.

         $P = \ldots$

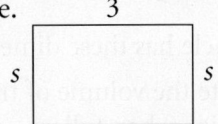

   (b) Rearrange the formula to make s the subject.

   (c) Use your rearranged formula to find the length s when
       the perimeter is 15 units.

3 This stem and leaf table shows the typing speeds in words per minute
of some people who took an audio-typing test.

```
        2  9
        3  2 2 4 7 8
        4  0 1 1 3 4 5 5 9
        5  1 3 4 4 6 7
        6  2
```

        Stem = 10 word per minute

   (a) How many people took the test?

   (b) What was the range of the typing speeds?

   (c) What was the median speed?

   (d) People with speeds of 38 words per minute or more had their work
      checked for accuracy. How many people was this?

4 Find the volume of each prism.
Give your answers in cm^3 correct to two significant figures.

(a)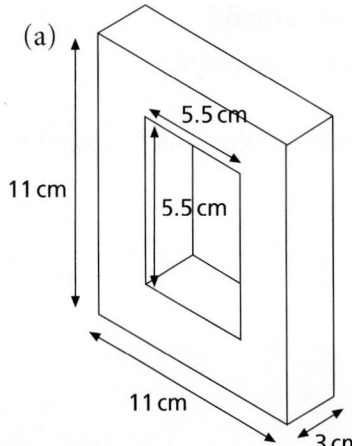

(b)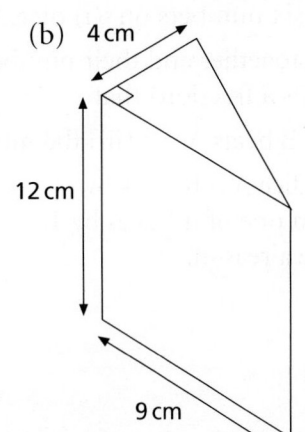

(c)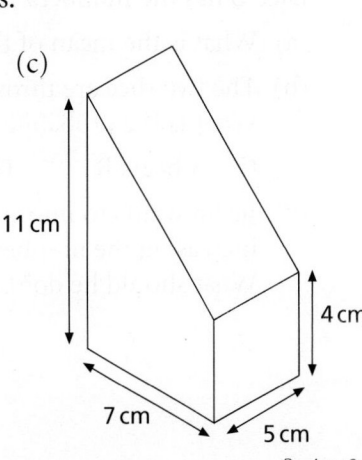

5 Solve the following equations.

  (a) $5a + 3(7 - a) = 7a - 4$ (b) $3(x - 1) - 4(x - 6) = 5(x - 3)$

6 (a) List all the outcomes when a 2p, 10p, 50p and £1coin are flipped.

  (b) What is the probability of all 4 coins showing a head?

  (c) What is the probability of all 4 coins showing the same face?

  (d) What is the probability of 3 or more coins showing a head?

  (e) What is the probability that there are more tails than heads showing?

7 A tin of treacle has these dimensions.

  (a) Calculate the volume of treacle
     it contains when full.

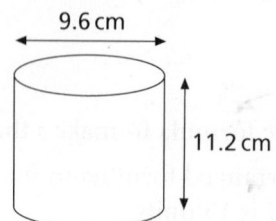

9.6 cm

11.2 cm

  (b) A full tin of treacle is accidentally spilled on a kitchen floor.
     The treacle spreads out until it is a puddle 3 mm thick.
     What is the area of the puddle in square centimetres?

8 A manufacturer keeps records of the number of fridges breaking down
during their one-year guarantee period.

Fridge model	Icefresh	Arctic chill	Coolmate
Number sold	57 897	236 745	186 772
Number breaking down	131	1047	550

  (a) For each fridge model, calculate, to 3 d.p., an estimate of the probability of
     a breakdown under guarantee.

  (b) On this basis, which model seems the most reliable?

9 Justin has two six-sided dice, A and B.
Dice A has the numbers 1, 1, 1, 2, 6, 6 on it.
Dice B has the numbers 1, 2, 2, 2, 2, 6 on it.

  (a) What is the mean of the six numbers on (i) dice A   (ii) dice B ?

  (b) The two dice are thrown together and their numbers are compared.
     What is the probability (as a fraction) that

     (i) A beats B   (ii) B beats A   (iii) the numbers are the same (a draw) ?

  (c) Justin wants to improve dice A's chances by
     increasing the number on one of its faces by 1.
     What should he do? Give a reason.